PHOTOGRAPHY FOUNDATIONS for art & design

PHOTOGRAPHY FOUNDATIONS for art & design

a guide to creative photography

second edition

mark galer

OXFORD AUCKLAND BOSTON JOHANNESBURG MELBOURNE NEW DELHI

Focal Press
An imprint of Butterworth-Heinemann
Linacre House, Jordan Hill, Oxford OX2 8DP
225 Wildwood Avenue, Woburn, MA 01801-2041
A division of Reed Educational and Professional Publishing Ltd

A member of the Reed Elsevier plc group

First published 1995
Reprinted 1997
Second edition 2000
Reprinted 2001

British Library Cataloguing in Publication Data
A catalogue record for this book is available from the British Library

Library of Congress Cataloguing in Publication Data
A catalogue record for this book is available from the Library of Congress

ISBN 0 240 51600 1

Printed and bound in Great Britain

Acknowledgements

Among the many people who helped make this book possible, I wish to express my gratitude to the following individuals:

Tom Davies for his enthusiasm for art and design education.
Jane Curry for her enthusiasm for photography in education.
Tim Daly for his advice and input to the first edition.
John Child and Adrian Davies for their input to the second edition.
Margaret Riley and Beth Howard for their vision and editorial input.
Canon Australia for their illustrative support.
The students of Spelthorne College, Photography Studies College and RMIT University for their overwhelming enthusiasm and friendship.
Dorothy, Matthew and Teagan for their love and understanding.

Picture Credits

Cover Design: Gus van der Heyde
Cover Illustration: Soek-Jin Lee

p.5 Gareth Neal; p.7 Eikoh Hosoe; p.8 Marc Riboud (Magnum Photos); p.9 Henri Cartier-Bresson (Magnum Photos); p.12 Burk Uzzle; p.13 Wil Gleeson; p.14 Joanne Arnold; p.15 Ian Berry (Magnum Photos); p.17 Tom Scicluna, Philip Leonard, Gareth Neal; p.18 Wil Gleeson; p.19 Ashley Dagg-Heston; p.21 Lorraine Watson; p.22 Kevin Ward; p.28 Sanjeev Lal, Tom Scicluna, Daniel Cox, Rew Mitchell; p.29 Renata Mikulik; p.30 Henri Cartier-Bresson; p.32 Jana Liebenstein; p.34 Chris Gannon; p.35 Alexandra Rycraft; p.40 Melanie Sykes, Clair Blenkinsop, Claire Ryder; p.41 Faye Gilding; p.43 Catherine Burgess; p.44 Tom Scicluna; p.45 Julia McBride; p.47 Claire Ryder, Daniel Shallcross; p.48 Matthew Houghton, Matthew Theobold; p.49 Chris Gannon; p.50 Darren Ware; p.51 Henry Peach-Robinson (The Royal Photographic Society); p.52 Peter Kennard; p.56 Angus McBean; p.57 David Hockney; p.59 Darren Ware, Zara Cronin; p.60 Lizette Bell; p.61 Ashley Dagg-Heston; p.64 Philip Budd; p.71 Gareth Neal; p.72 Gareth Neal, Paul Heath, Lynsey Berry; p.73 Mackenzie Charlton; p.76 Ansel Adams (CORBIS/Ansel Adams Publishing Rights Trust); p.77 Walker Evans; p.78 John Blakemore; p.79 Martin Parr (Magnum Photos); p.81 Michelle Greenhalgh; p.83 Matthew Orchard; p.85 Lucas Dawson, Alison Ward; p, 86 Lucas Dawson, Kalimna Brock; p.87 Mi-Ae Jeong; p.88 Matthew Orchard; p.89 Shaun Guest; p.90 Kata Bayer; p.92 Ann Ouchterlony; p.93 Stephen Rooke; p.94 Simon Sandlant; p.95 Sean Killen; p.97 Lizette Bell, Chris Augustnyk, Bec McCubbin; p.98 Shaun Guest, Lizette Bell; p.99 Andrew Goldie; p.100 Michael Mullan; p.101 Dorothea Lange; p.102 Kim Noakes; p.103 Michael Davies; p.106 Roly Imhoff; p.107 Sharounas Vaitkus; p.111 Michael Davies; p.112 Anthony Secatore; p.114 Burk Uzzle; p.117 Mike Wells; p.120 Arthur Sikeotis; p.159 Jana Liebenstein; p.167 Michael Wearne; p.174 Shaun Guest; p.175 Shaun Guest; p.176 Natalie Wright, Shane Bell, Georgia Tipperman.

All other photographs and illustrations by the author.

Contents

Advanced Study Guides

Technical Section

Introduction

This book is an introduction to photography for students studying Art & Design courses. The emphasis has been placed upon a creative rather than technical approach to the subject.

A structured learning approach

The photographic study guides contained in this book offer a structured learning approach that will give students a framework for working on design projects and the photographic skills for personal communication.
The study guides are intended as an independent learning resource to help build design skills, including the ability to research, plan and execute work in a systematic manner. Students are encouraged to adopt a thematic approach, recording all developmental work in the form of background work or study sheets.

Flexibility and motivation

The study guides contain a degree of flexibility in giving students the choice of subject matter. This allows the student to pursue individual interests whilst still directing their work towards answering specific design criteria. This approach gives the student maximum opportunity to develop self-motivation.

Implementation of the curriculum

The first three study guides are intended to be tackled sequentially and introduce no more technical information than is necessary for students to complete the work. This allows student confidence to grow quickly and enables less able students to complete all the tasks that have been set. The activities and assignments of the first three study guides provide the framework for the more complex assignments contained in the advanced section.
Each study guide requires approximately 25-40 hours to complete, plus additional time for independent research. '**Visual Literacy**' offers an optional photographic input for Media Studies students or design students studying photography in greater depth.

Teaching resources

In the '**Resources**' section of this book there is a work sheet and progress report which students can complete with the help of a teacher. This process will enable the student to organise their own efforts and gain valuable feedback about their strengths and weaknesses. The controlled test should be viewed as another assignment which the students can resource and then complete independently whilst being monitored. Students should try to demonstrate the skills which they have learnt in the preceding assignments.

Introduction to students

The study guides that you will be given on this course are designed to help you learn both the technical and creative aspects of photography. You will be asked to complete tasks including research activities and practical assignments. The information and experience that you gain will provide you with a framework for your future photographic work.

What is design?

Something that has been designed is something that has been carefully planned. When you are set a photographic design assignment you are being asked to think carefully about what you want to take a photograph of, what techniques you will use to take this image and what you want to say about the subject you have chosen. Only when the photograph communicates the information that you intended, is the design said to be successful.

By completing all the activities and assignments in each of the study guides you will learn how other images were designed and how to communicate visually with your own camera. You will be given the freedom to choose the subject of your photographs. The images that you produce will be a means of expressing your ideas and recording your observations.

Design is a process which can be learnt as a series of steps. Once you apply these simple steps to new assignments you will learn how to be creative with your camera and produce effective designs.

Using the study guides

The study guides have been designed to offer you support during your design work. On the first page of each study guide is a list of aims and objectives laying out the skills covered and how they can be achieved.

The activities are to be undertaken after you have first read and understood the supporting section on the same page. If at any time you feel unclear about what is being asked of you, consult a teacher.

Equipment needed

The course that you are following has been designed to teach you photography with the minimum amount of equipment. You will need a 35mm SLR camera with manual controls or manual override if automatic. Consult your teacher or a photographic specialist store if you are in doubt. Many dealers can supply second-hand equipment complete with a guarantee at reasonable prices. Large amounts of expensive equipment will not make you a better photographer. Many of the best professional photographers use less equipment than some amateurs. There are some areas of photography, however, which do require some very specialist equipment. These include some areas of sport and wildlife photography where you are unable to get very close to your subject. If these areas are of particular interest, you will need to think carefully about which of these sporting activities or animals are possible with the equipment you intend to use.

Research and resources

The way to get the best out of each assignment is to use the activities contained in the study guides as a starting point for your research.

You will only realise your full creative potential by looking at a variety of images from different sources. Creative artists and designers find inspiration for their work in different ways, but most find that they are influenced by other work they have seen and admired.

> 'The best designers are those who have access to the most information.'
>
> Stephen Bailey - former director of the Design Museum.

Getting started

Start by collecting and photocopying images that are relevant to the activity you have been asked to complete. This collection of images will act as a valuable resource for your future work. By taking different elements from these different images, e.g. the lighting technique from one and the vantage point from another, you are not copying somebody else's work but using them as inspiration for your own creation.

Talking through ideas with other students, friends, members of your family and with a teacher will help you to clarify your thinking, and develop your ideas further.

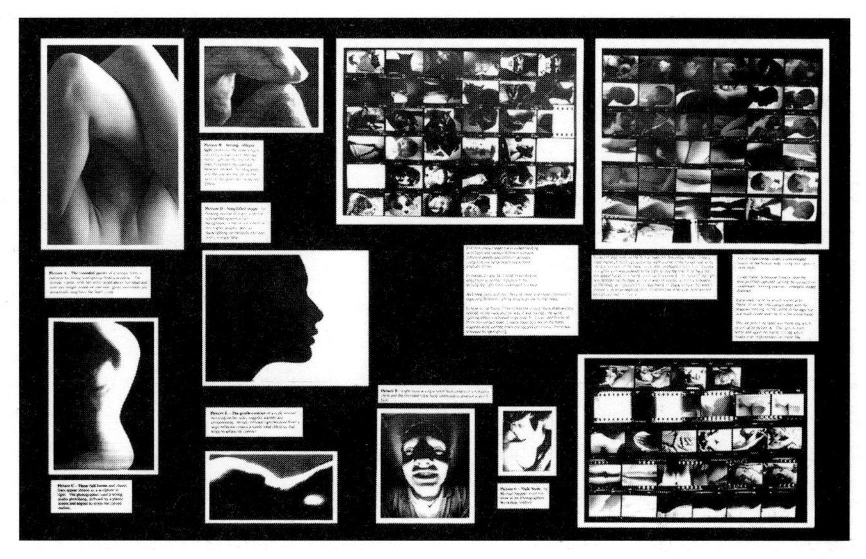

Student study sheet

Choosing resources

When you are looking for images that will help you with your research activities try to be very selective, using high quality sources. Not all photographs that are printed are necessarily well designed or appropriate to use. Good sources of photography may include high quality magazines and journals, photographic books and photography exhibitions. You may have to try many libraries to find appropriate material. Keep an eye on the local press to find out what exhibitions are coming to your local galleries.

Presentation of research

In each assignment you are asked to provide evidence of how you have developed your ideas and perfected the techniques you have been using. This should be presented neatly and in an organised way so that somebody assessing your work can easily see the creative development of the finished piece of work.

You should edit your proof sheet including any alterations to the original framing with a chinagraph pencil or indelible marker pen. Make comments about these images to show how you have been selective and how this has influenced subsequent films that you have taken. You should clearly state what you were trying to do with each picture and comment on its success. You should also state clearly how any theme which is present in your work has developed.

Make brief comments about images that you have been looking at and how they have influenced your own work. Photocopy these images if possible and include them with your research.

All contact prints and photographs should be easily referenced to relevant comments using either numbers or letters as a means of identification. This coding will insure that the person assessing the work can quickly relate the text with the image that you are referring to.

Research and all proof sheets should be carefully mounted on card for display.

Presentation of finished work

The way you present your work can influence your final mark. Design does not finish with the print. Try laying out the work before mounting it on your card. Use rulers or a straight edge in aligning work if this is appropriate. Make sure the prints are neatly trimmed and that any writing has been spell-checked and is grammatically correct.

Final work should be mounted on card using a suitable adhesive. Adhesives designed to stick paper do not work efficiently on resin-coated photographic paper. Photographic prints are normally either dry mounted using adhesive tissue and a dry mounting press or window mounted. Both are time consuming and require a fair amount of skill. A cheaper and quicker alternative is to use double-sided tape applied to each corner of the photograph.

If you choose to write a title on the front of the sheet it is advisable that you either use a lettering stencil or generate the type using a computer. Be sure to write your name and project title on the back of this card so the person assessing the work can return the work to you quickly.

Storage of work

Assignment work should be kept clean and dry, preferably using a folder slightly larger than the size of your finished sheets. It is recommended that you standardise your presentation so that your final portfolio looks neat and presentable.

Negatives should always be stored in negative file sheets in dry dust-free environments that will ensure clean reprints can be made if necessary.

The Frame

Pavement - Gareth Neal

aims

- To develop an awareness of how a photographic print is a two-dimensional composition of lines, shapes and patterns.
- To develop an understanding of how different ways of framing can affect both the emphasis and the meaning of the subject matter.

objectives

- **Research** - look at the composition of different photographs and the design techniques used. Record these observations and findings.
- **Analyse and evaluate critically** - exchange your ideas and opinions with others.
- **Develop ideas** - produce a study sheet that documents the progress and development of your own ideas.
- **Personal response** - produce and present photographic prints through close observation and selection that demonstrate how the frame can create compositions of shape, line and pattern and a personal theme.

Introduction

From photographs each of us can learn more about the world. Images not only inform us about the products we never knew we needed, the events, people and places too distant or remote for us to see with our own eyes, but also tell us more about the things we thought we already knew.

Most of us are too preoccupied to stand and look at something for any great amount of time. We glance at something briefly and think we have seen it. Our conditioning or desires often tell us what we have seen or would like to see. When we look at a photograph of something ordinary, however, it may show us the object as we had never seen it before. With a little creative imagination and a little photographic technique it is possible to release the extraordinary from the ordinary.

Bill Brandt in 1948 said that 'it is the photographer's job to see more intensely than most people do. He must keep in him something of the child who looks at the world for the first time or of the traveller who enters a strange country.'

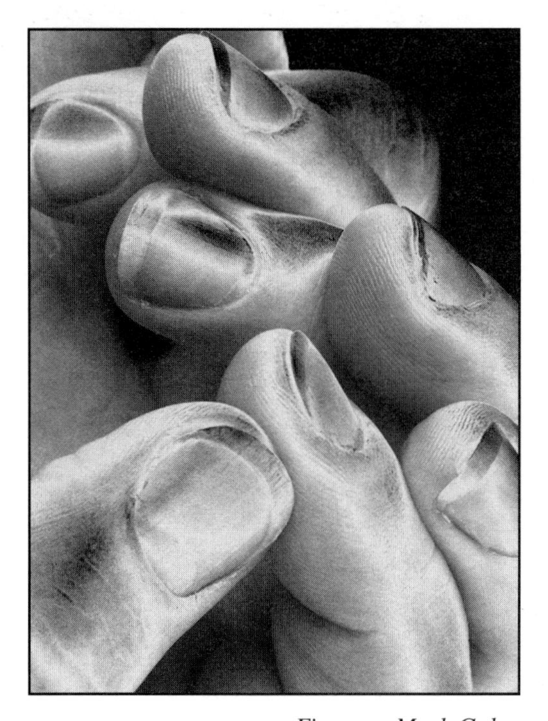

Fingers - Mark Galer

Choosing a subject

In order to photograph something that will be of interest to others you must first remove the blinkers and photograph something that is of interest to you. Your first creative decision is an important one. What will you choose to photograph? Your first technical decision is how to frame it.

Composition

Framing the subject

The photographer Robert Capa said 'If your pictures aren't good, you're not close enough.' He was saying that the subject matter can look unimportant and not worthy of closer attention. There is also a danger that the photographer will not have control over the composition.
A common mistake made by many amateur photographers is that they stand too far away from their subject matter, in a desire to include everything. Their photographs become busy, unstructured and cluttered with unwanted detail which distracts from the primary subject matter.

Embrace - Eikoh Hosoe 1970

The photograph above is a study of the human figure and also a composition of shape, tone and line. There are three dominant shapes. The woman's leg, the man's back and also the third shape which is created between the frame and the man's back. The act of framing a subject using the viewfinder of the camera imposes an edge that does not exist in reality. This frame also dissects familiar objects to create new shapes. The shapes that this frame creates must be studied carefully in order to create successful compositions.
The powerful arc of the man's back is positioned carefully in relationship to the edge of the frame and the leg of the woman is added to balance the composition.

Filling the frame

When the photographer moves closer, distracting background can be reduced or eliminated. There are less visual elements that have to be arranged and the photographer has much more control over the composition. Many amateurs are afraid of chopping off the top of someone's head or missing out some detail that they feel is important. Unless the photograph is to act as a factual record the need to include everything is unnecessary.

Father and Child, Vietnam - Marc Riboud

When the viewer is shown a photograph they have no way of knowing for sure what lies beyond the frame. We often make decisions on what the photograph is about from the information we can see. We often have no way of knowing whether these assumptions are correct or incorrect.

The photograph above is of a father and child. The protective hands of a father figure provide the only information most people need to arrive at this conclusion. In order to clarify any doubt the photographer may have decided to move further back to include the whole figure. The disadvantage in doing this would have been that the background would also begin to play a large part in the composition and the power of this portrait of a child and his father would have been lost. Photographers do have the option, however, of taking more than one photograph to tell a story.

Activity 1

Look through assorted photographic books and observe how many photographers have moved in very close to their subjects. By employing this technique the photographer is said to 'fill the frame' and make their photographs more dramatic.

Find two examples of how photographers seek simple backgrounds to remove unwanted detail and to help keep the emphasis or 'focal point' on the subject.

The whole truth?

Photographs provide us with factual information but sometimes we do not have enough information to be sure what the photograph is about.

Hyde Park - Henri Cartier-Bresson

What do we know about this old lady or her life other than what we can see in the photograph? Can we assume she is lonely as nobody else appears within the frame? Could the photographer have excluded her grandchildren playing close at hand to improve the composition or alter the meaning? Because we are unable to see the event or subject that the photograph originated from we are seeing it out of context.

Activity 2

Read the following passage taken from the book *The Photographer's Eye* by John Szarkowski and answer the questions below.

> 'To quote out of context is the essence of the photographer's craft. His central problem is a simple one: what shall he include, what shall he reject? The line of decision between in and out is the picture's edge. While the draughtsman starts with the middle of the sheet, the photographer starts with the frame.
> The photograph's edge defines content. It isolates unexpected juxtapositions. By surrounding two facts, it creates a relationship. The edge of the photograph dissects familiar forms, and shows their unfamiliar fragment. It creates the shapes that surround objects.
> The photographer edits the meanings and the patterns of the world through an imaginary frame. This frame is the beginning of his picture's geometry. It is to the photograph as the cushion is to the billiard table.'

Q. What does John Szarkowski mean when he says that photographers are quoting 'out of context' when they make photographic pictures?

Q. The frame often 'dissects familiar forms'. At the end of the last century photography was having a major impact on Art. Impressionist artists such as Degas were influenced by what they saw.
Find an example of his work which clearly shows this influence and explain why the public might have been shocked to see such paintings.

Subject placement

When the photographer has chosen a subject to photograph there is often the temptation, especially for the untrained eye, to place the subject in the middle of the picture without considering the overall arrangement of shapes within the frame. If the focal point is placed in the centre of the frame, the viewers eye may not move around the whole image and this often leads to a static and uninteresting composition. The design photographer should think carefully where the main subject is placed within the image, only choosing the central location after much consideration.

> 'A picture is well composed if its constituents - whether figures or apples or just shapes - form a harmony which pleases the eye when regarded as two-dimensional shapes on a flat ground.'
>
> Peter and Linda Murray - *A Dictionary of Art and Artists*

The rule of thirds

Rules of composition have been formulated to aid designers create harmonious images which are pleasing to the eye. The most common of these rules are the '**golden section**' and the '**rule of thirds**'.

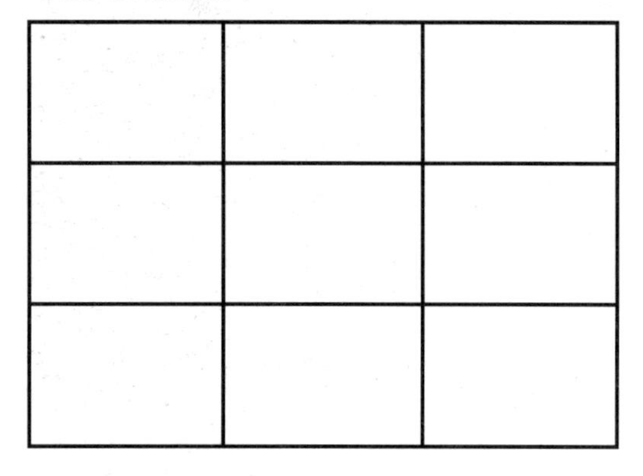

The rule of thirds

The golden section is the name given to a traditional system of dividing the frame into unequal parts which dates back to the time of Ancient Greece. The rule of thirds is the simplified modern equivalent. Try to visualise the viewfinder as having a grid which divides the frame into three equal segments, both vertically and horizontally. Many photographers and artists use these lines and their intersection points as key positions to place significant elements within the picture.

Breaking the rules

Designers who are aware of these rules often break them by deliberately placing the elements of the image closer to the edges of the frame. This can often be effective in creating '**dynamic tension**' where a more formal design is not needed.

Activity 3

Find two examples of photographs that follow the rule of thirds and two examples that do not. Comment briefly on why and how you think the composition works.

Balance

In addition to content a variety of visual elements such as line, colour and tone often influence a photographer's framing of an image. The eye naturally or intuitively seeks to create a '**symmetry**' or a harmonious relationship between these elements within the frame. When this is achieved the image is said to have a sense of '**balance**'. The most dominant element of balance is visual weight created by the distribution of light and dark tones within the frame. To frame a large dark tone on one side of the image and not seek to place tones of equal visual weight on the other side will create imbalance in the image. An image that is not balanced may appear heavy on one side. Visual tension is created within an image that is not balanced. Balance, although calming to the eye, is not always necessary to create an effective image. Communication of harmony or tension is the deciding factor of whether balance is desirable in the image.

Tai Chi

Activity 4

Collect one image where the photographer has placed the main subject off-centre and retained a sense of balance and one image where the photographer has placed the main subject off-centre and created a sense of imbalance.

Discuss the possible intentions of the photographer in creating each image.

Create four images, placing the focal point and/or visual weight in different areas of the frame.

Discuss whether each image is balanced.

Line

The use of line is a major design tool that the photographer can use to structure the image. Line becomes apparent when the contrast between light and dark, colour, texture or shape serves to define an edge. The eye will instinctively follow a line. Line in a photograph can be described by its length and angle in relation to the frame (itself constructed from lines).

Horizontal and vertical lines

Horizontal lines are easily read as we scan images from left to right comfortably. The horizon line is often the most dominant line within the photographic image. Horizontal lines within the image give the viewer a feeling of calm, stability and weight. The photographer must usually be careful to align a strong horizontal line with the edge of the frame. A sloping horizon line is usually immediately detectable by the viewer and the feeling of stability is lost.

Vertical lines can express strength and power. This attribute is again dependent on careful alignment with the edge of the frame. This strength is lost when the camera is tilted to accommodate information above or below eye level. The action of '**perspective**' causes parallel vertical lines to lean inwards as they recede into the distance.

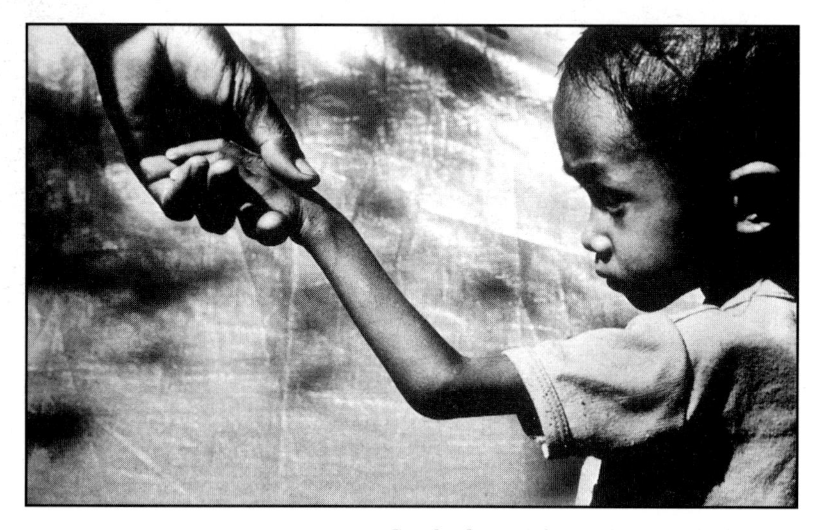

Cambodian Refugee Camp - Burk Uzzle

Suggested and broken line

Line can be designed to flow through an image. Once the eye is moving it will pick up a direction of travel and move between points of interest. The photograph above is a good example of how the eye can move through an image. Viewers tend to look briefly at the adult hand holding the child's, before quickly moving down the arm to the face. The direction of the child's gaze returns the viewer's attention to the hands. A simple background without distracting detail helps to keep our attention firmly fixed on this relationship.

Diagonal lines

Whether real or suggested, these lines are more dynamic than horizontal or vertical lines. Whereas horizontal and vertical lines are stable, diagonal lines are seen as unstable (as if they are falling over) thus setting up a dynamic tension or sense of movement within the picture.

Daimaru

Wil Gleeson

Curves

A curved line is very useful in drawing the viewer's eye through the image in an orderly way. The viewer often starts viewing the image at the top left-hand corner and many curves exploit this. Curves can be visually dynamic when the arc of the curve comes close to the edge of the frame or directs the eye out of the image.

Activity 5

Find two examples of photographs that use straight lines as an important feature in constructing the pictures' composition.
Find one example where the dominant line is either an arc or S-curve.
Comment briefly on the contribution of line to the composition of each example.
Construct an image where the viewer is encouraged to navigate the image by the use of suggested line and broken line between different points of interest.

Vantage point

A carefully chosen viewpoint or '**vantage point**' can often reveal the subject as familiar and yet strange. In designing an effective photograph that will encourage the viewer to look more closely, and for longer, it is important to study your subject matter from all angles. The 'usual' or ordinary is often disregarded as having been 'seen before' so it is sometimes important to look for a fresh angle on a subject that will tell the viewer something new.

The Beach - Joanne Arnold

When we move further away from our subject matter we can start to introduce unwanted details into the frame that begin to detract from the main subject. Eventually the frame is so cluttered that it can look unstructured. The careful use of vantage point can sometimes overcome this. A high or low vantage point will sometimes enable the photographer to remove unwanted subject matter using the ground or the sky as an empty backdrop.

Activity 6

Find two examples of photographs where the photographer has used a different vantage point to improve the composition.
Comment on how this was achieved and how this has possibly improved the composition.

Depth

When we view a flat two-dimensional print which is a representation of a three-dimensional scene, we can often recreate this sense of depth in our mind's eye. Using any perspective present in the image and the scale of known objects we view the image as if it exists in layers at differing distances. Successful compositions often make use of this sense of depth by strategically placing points of interest in the foreground, the middle distance and the distance. Our eye can be led through such a composition as if we were walking through the photograph observing the points of interest on the way.

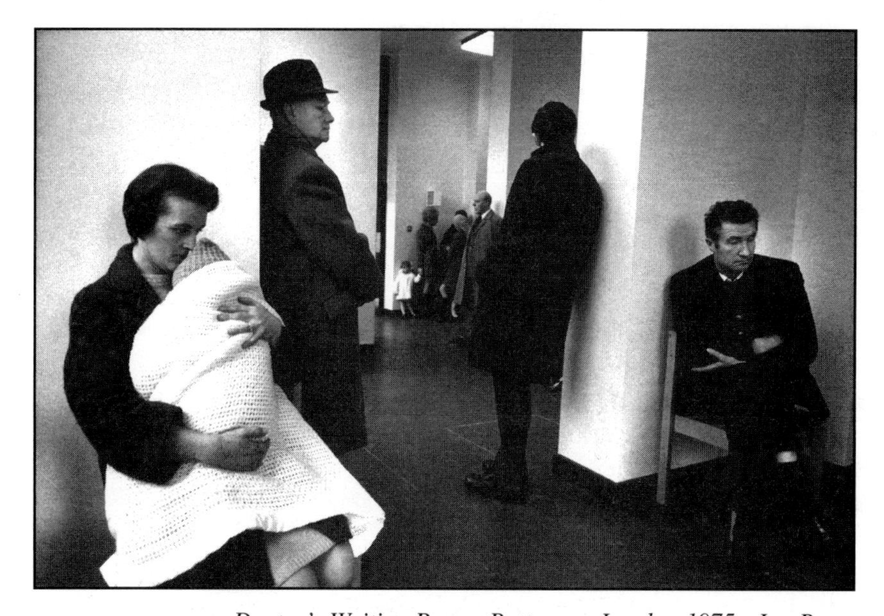

Doctor's Waiting Room, Battersea, London 1975 - Ian Berry

In the image above our eyes are first drawn to the largest figures occupying the foreground on either side of the central doorway. In a desire to learn more from the image our eyes quickly progress towards the figures occupying the middle distance. Appearing as lazy sentinels the figures lean against the doorway and move our gaze towards the focal point of the photograph, the small girl holding her mother's hand in the centre of the image.
The technique of drawing us into the photograph is used in many photographs and can be also be exploited using dark foreground tones drawing us towards lighter distant tones.

Activity 7

Find two photographs where the photographer has placed subject matter in the foreground, the middle distance and the distance in an attempt either to fill the frame or to draw our gaze into the image.
Comment briefly on where you feel the focal points of these images are.

Practical assignment

Produce a set of six photographs investigating natural or man-made forms. Your work should demonstrate how the frame can be used to create compositions of shape, line and pattern. You should consider not only the shapes of your subject matter but also those formed between the subject and the frame.

Choosing a theme

Your photographs should develop a clearly defined theme. This could be several different ways of looking at one subject or different subjects that share something in common, e.g. a similar pattern or composition. If you are unfamiliar with your camera choose a subject that will keep still, allowing you time to design the composition.
A possible title for your set of prints could be:

1. Patterns in nature.
2. Rhythms of life.
3. Urban patterns.

Your work should:

a) make use of differences in subject distance including some work at, or near, the closest focusing distance of your camera lens;
b) show that you have considered the rule of thirds;
c) demonstrate the creative use of line in developing your compositions;
d) make use of different vantage points;
e) show that you have thought carefully about the background and the foreground in making your composition.

Note. Implement aspects of your research during your practical assignment.

Resources

Basic Photography - Michael Langford. Focal Press. London. 1997.
Black and White Photography - Rand/Litschel. West Publishing Co. Minneapolis. 1994.
Photography Composition - Tom Grill and Mark Scanlon. Fountain Press Ltd. 1984
Photography - London and Upton. Harper Collins College Publishers. New York. 1994.
The Image - Michael Freeman. Collins Photography Workshop.
The Photographer's Eye - John Szarkowski. Museum of Modern Art. New York. 1966.

Spiral Staircase - Tom Scicluna

Many photographers use the lines and intersection points of the rule of thirds as key positions to place significant elements in the picture.

Arc - Tom Scicluna

By tilting the camera the shapes and lines can be carefully organised within the frame. The dominant arc in this image sweeps the edge of the frame and exits at the bottom right-hand corner.

Staircase - Philip Leonard

Diagonal lines that appear in a picture, whether real or suggested, are more dynamic than horizontal lines. The diagonal lines have been arranged to enter the corners of this photograph.

Stacked Chairs - Gareth Neal

A carefully chosen viewpoint can often reveal the subject as familiar and yet strange. A student has explored the interesting lines created by a stack of chairs.

Gallery

Wil Gleeson

Light

Hands - Ashley Dagg-Heston

aims

- To develop knowledge and understanding of how the quality and direction of light can change character and mood.
- To develop an awareness to the limitations of film in recording subject contrast.

objectives

- **Research** - produce a study sheet that looks at the 'atmospheric' lighting of several different photographs and the lighting techniques employed by the photographers.
- **Analyse and evaluate** - the effectiveness of the work you are studying and exchange ideas and opinions with other students.
- **Discussion** - exchange ideas and opinions with other students.
- **Personal response** - produce images through close observation and selection that demonstrate how the quality of light and its direction affects communication.

Introduction

Light is the black and white photographer's medium. The word photography is derived from the ancient Greek words photos and graph meaning 'light writing'. In photography, shadows, reflections, patterns of light, even the light source itself may become the main subject and solid objects may become incidental to the theme.

Projection - Student photograph

Creating atmosphere

Directional light from the side and/or from behind a subject can produce some of the most evocative and atmospheric black and white photographs. Most snapshots by amateurs, however, are taken either outside when the sun is high or inside with a flash mounted on the camera. Both these situations give a very flat and even light which may be ideal for some colour photography but for black and white photography it all too often produces grey, dull and uninteresting photographs.

Learning to control light and use it creatively is an essential skill for a good photographer. When studying a photograph that has been well lit you need to make three important observations concerning the use of light:

1. What type or quality of light is being used?
2. Where is it coming from?
3. What effect does this light have upon the subject and background?

Quality of light

Light coming from a compact source such as a light bulb or the sun can be described as having a very 'hard quality'. The shadows created by this type of light are dark and have well-defined edges.

Light coming from a large source such as sunlight that has been diffused by clouds or a light that has been reflected off a large bright surface is said to have a very 'soft quality'. The shadows are less dark (detail can be seen in them) and the edges are not clearly defined.

The smaller the light source, the harder the light appears.
The larger the light source, the softer the light appears.

Claire - Lorraine Watson

A soft light positioned high and to the right gives a flattering and glamourous effect. Note the soft shadows from the glasses.

Claire - Lorraine Watson

A single hard light positioned directly above casts deep shadows around the eyes giving a dramatic effect which complements the pose. Note how the hands add to the expression.

Activity 1

Look through assorted photographic books and find some examples of subjects lit by hard light and examples of subjects lit by soft light.

Describe the effect the light has on the subjects' texture, form and detail and the overall mood of the picture.

Direction of light

The direction of light decides where the shadows will fall and its source can be described by its relative position to the subject. The light may be high, low, to one side, in front of or behind the subject.
The subject may be lit by a single light source or more than one. This additional light may be reflected back onto the subject from a nearby surface or may be shining directly onto the subject from a second light or be used to illuminate the background.

Broken Light - Kevin Ward

Activity 2

Find an example of a photograph where the subject has been lit by a single light source and an example where more than one light has been used.
Describe in each the quality and position of the brightest or main light and the effect this has on the subject. In the second example describe the quality and effect the additional light has.

Subject contrast

Contrast is the degree of difference between the lightest and darkest tones of the subject or photographic image. A high-contrast photograph is where the dark and light tones dominate over the mid-tones within the image. The highest contrast image possible is one that contains only two tones, black and white, and where no mid-tones remain. A low-contrast image is one where mid-tones dominate the image and there are few if any tones approaching black or white.

Each subject framed by the photographer will include a range of tones from dark to light. The combined effects of quality of light and lighting direction give the subject its contrast. It is contrast that gives dimension, shape and form to the subject photographed.

Contrast and the limitations of film

The human eye can register detail in a wide range of tones simultaneously. Film is unable to do this. It can record only a small range of what human vision is capable of seeing.

High contrast

Low contrast

Cloud cover diffuses and softens the light leading to lower contrast. Shadows appear less harsh and with softer edges. The lighting may be described as being flat and the film will be able to record all of the tones from black to white if the exposure is correct. When harsh directional light strikes the subject the overall contrast of the scene increases. The photographer may now be able to record only a small selection of the broad range tones.

Increasing exposure will reveal more detail in the shadows and dark tones.
Decreasing exposure will reveal more detail in the highlights and bright tones.

If the photographer wishes to photograph in harsh directional sunlight it is appropriate to increase the exposure from that indicated by the camera's TTL light meter to avoid underexposure. An additional one stop is usually sufficient to protect the shadow detail on the film.

If a subject is lit by harsh directional light increase the exposure by one stop from that indicated by the TTL light meter.

Exposure compensation

When you take a light meter reading of a subject you are taking an average reading between the light and the dark tones you have framed. The meter reading is accurate when there is an even distribution of tones, or the dominant tone is neither dark nor light. It is very important that you '**compensate**' or adjust the exposure when the framed area is influenced by a tone that is dark, light or very bright.
If your subject is in front of a bright light, such as a window or the sky, the light meter will indicate an average reading between your subject and the very bright tone. The camera's meter will be influenced by this tone and indicate an exposure setting that will reduce the light reaching the film. In these situations you have to override the meter and increase the exposure to avoid underexposing the subject.

The Rhondda - Mark Galer

Most cameras take information for the light reading mainly from the centre of the viewfinder. When you need to set the exposure for a subject that requires compensation you can:

1. Move in close so that your chosen subject fills the frame and set the exposure. Move back to your chosen camera position and take the shot at the same setting.
2. Point the centre of the viewfinder away from the bright light, set the exposure and reposition the camera. Some cameras have a memory lock to help you do this.

Activity 3

Find two photographs where the photographer has shot into the light or included the light source.
Explain how the photographer may have gone about taking a light meter reading for these photographs.

Depth of field

You can increase or decrease the amount of light reaching the film by moving one of two controls. By changing the shutter speed (the amount of time the shutter stays open for) or by changing the f-stop (the size of the aperture through the lens).
If you change the aperture, the final appearance of the photograph can differ greatly. This will be the area of sharp focus in the scene, from the nearest point that is sharp to the farthest. This is called '**depth of field**'.

> **The widest apertures (f2, f4) give the least depth of field.**
> **The smallest apertures (f11, f16) give the most depth of field.**

Try this simple experiment to help you understand depth of field. Look at your hand at arm's length in front of your eyes. You will notice the room behind your hand is out of focus. Similarly if you look at the far wall of the room, your hand will appear out of focus. This is called 'shallow depth of field' and is how your camera will see it at the widest aperture. If you progressively reduce the aperture value, known as stopping down, the amount that is in focus gradually increases until you reach f16 or f22 when you achieve 'maximum depth of field'. At this aperture it is possible to have both your hand and the wall in focus at the same time.
When you look through your camera and adjust the aperture no change appears to take place. This is because the lens stays at its widest aperture to allow a bright viewfinder image and easy focusing. The instant you depress the shutter the aperture closes to the f-stop you have selected. Some cameras have a 'depth of field preview' located near the lens which allows you to see the true extent of sharpness.

f16

f8

f4

Depth of field

Activity 4

Find four examples of photographs which make use of maximum depth of field and examples which have very shallow depth of field.
Describe how the photographer's selective use of aperture affects the subject in each of the photographs you have chosen.

Basic studio lighting

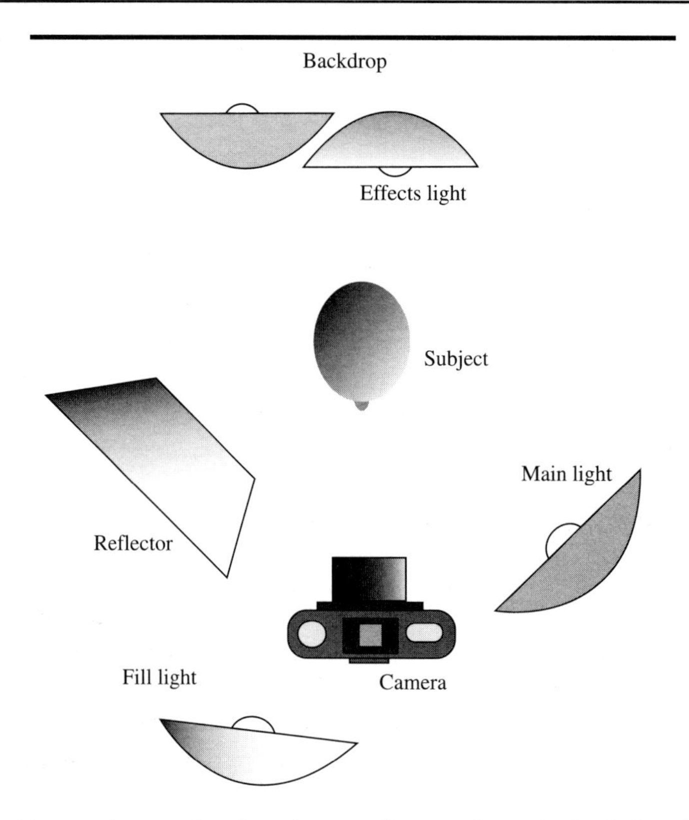

The subject - this may have to be placed some distance from the backdrop if the subject and backdrop are to be illuminated separately.

Main light source - the position is optional depending upon the desired effect required. If deep shadows are created by the main light they can be softened in one of three ways:

1. Reflector - this is used to bounce light from the main light back into the shadow areas.
2. Fill - this is usually a weaker light or one moved further away and is usually positioned by the camera.
3. Diffusion - the hard directional light can be softened at the source either by bouncing it off a white surface or by using tracing paper placed in front of the light itself.

Effects light - this can be shone onto the background to create 'tonal interchange'. This is a technique where the photographer places dark areas of the subject against light areas of the background and vice versa.
Alternatively the effects light can be shone directly onto the back of the subject so as to 'rim light' otherwise dark areas. Care must be taken to avoid shining the light directly into the camera lens. This is usually achieved by placing the effects light low or obscuring the light directly behind the subject itself.

Practical assignment

Produce a set of six prints that express your feelings towards one of the following titles:

1. Shadows and silhouettes.
2. Broken light.
3. The twilight hours.
4. Face and figure.

Your final work should develop the insight gained from all areas of your research and should include investigations into both natural and artificial light. Try experimenting with window light, studio lights, projected images, light bulbs, candles, torches etc.
If you are having difficulty thinking of a suitable subject try working with another student using each other as 'models'. You may decide to use some other person for the final piece of finished work.

Presentation of work

Research and all contact prints should be carefully mounted on large sheets of paper or thin card. Explanatory notes and comments should be made directly onto this sheet or on paper and glued into place. You should edit your proof sheet including any alterations to the original framing with a chinagraph pencil or indelible marker pen. You should clearly state what you were trying to do with each picture and comment on its success. Proof sheets and photographs should be easily referenced to relevant comments using either numbers or letters as a means of identification. You should clearly state how your theme has developed and what you have learnt from your research and how this has contributed towards your final set of prints.

Resources

Basic Photography - Michael Langford. Focal Press. London. 1997.
Black and White Photography - Rand/Litschel. West Publishing Co. Minneapolis. 1994.
Photography Composition - Tom Grill and Mark Scanlon. Fountain Press Ltd. 1984.
Photography - London and Upton. Harper Collins College Publishers. New York. 1994.
Photographic Lighting - Child/Galer. Focal Press. London. 1999.

Hands 1 - Sanjeev Lal

Using a single hard light source the student has taken care to frame the subject in a dynamic way. Note how the hands add to the expression of the portrait.

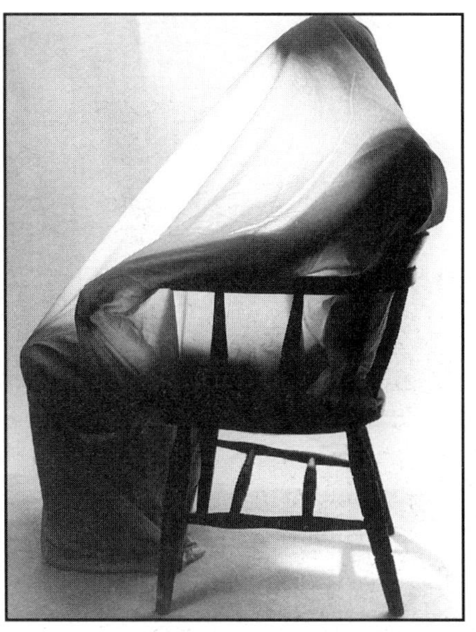

Furniture Constructions - Daniel Cox

A single light pointing back towards the camera has been placed behind the figure sitting in the chair. The student has rearranged the order of the chair frame, its cover and the sitter.

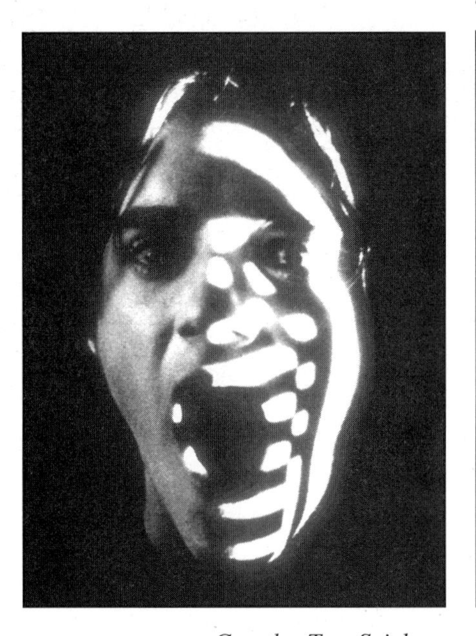

Gareth - Tom Scicluna

The contours of the face break the pattern of light being shone from a slide projector positioned to the right of the face.

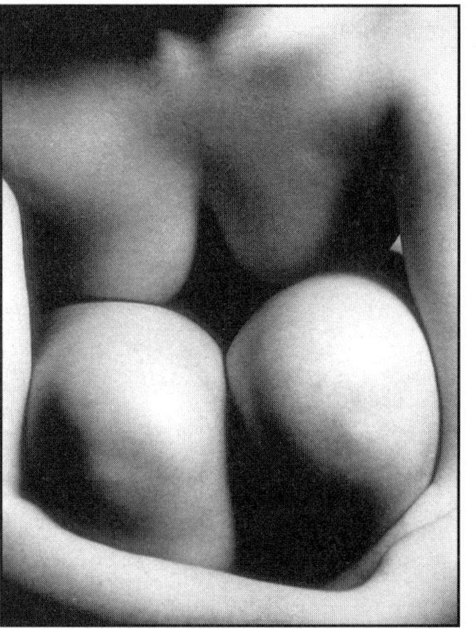

Embrace - Rew Mitchell

Hard light, positioned above and left, is used to create this dramatic study of the human figure. The arms are used to create a frame within a frame.

Time

Movement - Renata Mikulik

aims

- To develop an understanding of how a photograph can describe a subject over a period of time as selected by the photographer.
- To develop an appreciation of how the selected period of time or 'shutter speed' can affect the visual outcome of the print.

objectives

- **Research** - produce a study sheet that looks at the effect that varying the shutter speed has on a moving subject.
- **Analyse and evaluate critically** - exchange ideas and opinions with other students on the work you are studying.
- **Develop ideas** - produce a study sheet that documents the progress and development of your own ideas
- **Personal response** - produce a series of photographic prints that explore a moving subject by using a variety of techniques and shutter speeds.

Introduction

All photographs are time exposures of shorter or longer duration, and each describes an individually distinct parcel of time. The photographer, by choosing the length of exposure, is capable of exploring moving subjects in a variety of ways.
By choosing long exposures moving objects will record as blurs. This effect is used to convey the impression or feeling of motion. Although describing the feeling of the subject in motion much of the information about the subject is sacrificed to effect. By selecting fast shutter speeds photographers can freeze movement. We can see the nature of an object in motion, at a particular moment in time, that the human eye is unable to isolate.

Greek Girl - Henri Cartier-Bresson

The decisive moment

Henri Cartier-Bresson in 1954 described the visual climax to a scene which the photographer captures as being the 'decisive moment'.
In the flux of movement a photographer can sometimes intuitively feel when the changing forms and patterns achieve balance, clarity and order and when the image becomes, for an instant, a picture.

Activity 1

Look at a Henri Cartier-Bresson photograph and discuss why you think that capturing the decisive moment has added to the picture's quality.

Fast shutter speeds

By freezing thin slices of time, it is possible to explore the beauty of form in motion. A fast shutter speed may freeze a moving subject yet leave others still blurred. The ability to freeze subject matter is dependent on its speed and the angle of movement in relation to the camera. For subject matter travelling across the camera's field of view, relatively fast shutter speeds are required, compared to the shutter speeds required to freeze the same subject travelling towards or away from the camera.

Songkran Festival

Limitations of equipment

Wide apertures in combination with bright ambient light and/or fast film allow the use of fast shutter speeds in order to freeze rapidly moving subject matter. Some telephoto and zoom lenses only open up to f4 or f5.6. If used with a slow or medium-speed film there is usually insufficient light to use the fastest shutter speeds available on the camera.

Activity 2

Find an example of a photograph where the photographer has used a very fast shutter speed and describe the subject matter including the background. Discuss any technical difficulties the photographer may have encountered and how he or she may have overcome them.
Discuss what has happened to the depth of field and why.
Discuss whether the image gives you the feeling of movement stating the reasons for your conclusion.

Panning

Photographers can follow the moving subject with the camera in order to keep the subject within the frame. This technique called '**panning**' allows the photographer to use a slower shutter speed than would otherwise have been required if the camera had been static. The ambient light is often insufficient to use the very fast shutter speeds making panning essential in many instances.

For successful panning the photographer must aim to track the subject before the shutter is released and follow through or continue to pan once the exposure has been made. The action should be one fluid movement without pausing to release the shutter. A successful pan may not provide adequate sharpness if the focus is not precise.

In order to have precise focusing with a moving subject the photographer may need to use a fast predictive autofocus system or pre-focus on a location which the moving subject will pass through.

Bird in Flight - Jana Liebenstein

Activity 3

Take four images of a running or jumping figure using fast shutter speeds (faster than 1/250 second). Vary the direction of travel in relation to the camera and attempt to fill the frame with the figure. Examine the image for any movement blur and discuss the focusing technique used.

Take four images of the same moving subject using shutter speeds between 1/15 and 1/125 second. Pan the camera to follow the movement. The primary subject should again fill the frame. Discuss the visual effect of each image.

Slow shutter speeds

When the shutter speed is slowed down movement is no longer frozen but records as a streak across the film. This is called '**movement blur**'. By using shutter speeds slower than those normally recommended for use with the lens, movement blur can be created with relatively slow moving subject matter. Speeds of 1/30, 1/15, 1/8 and 1/4 second can be used to create blur with a standard lens. If these slow shutter speeds are used and the camera is on a tripod the background will be sharp and the moving subject blurred. If the camera is panned successfully with the moving subject the background will provide most of the blur in the form of a streaking effect in the direction of the pan.

Daimaru

Camera shake

Movement blur may also be picked up from camera movement as a result of small vibrations transmitted to the camera from the photographer's hands. This is called '**camera shake**'. To avoid camera shake a shutter speed roughly equal to the focal length of the lens is usually recommended, e.g. 1/30 second at 28mm, 1/60 second at 50mm and 1/125 at 135mm. Many cameras give an audible signal when shutter speeds likely to give camera shake are being used.

With careful bracing, slower speeds than those recommended can be used with great success. When using slow shutter speeds the photographer can rest elbows on a nearby solid surface, breathe gently and release the shutter with a gentle squeeze rather than a stabbing action.

Activity 4

Find a photograph where the photographer has used a slow shutter speed and describe the subject matter including the background. Discuss any technical difficulties the photographer may have encountered and how these may have been overcome.

Discuss what has happened to the depth of field and why.

Discuss whether the image gives you the feeling of movement stating the reasons for your conclusions.

Photographic techniques

Very long exposures

For this technique you will need to mount the camera on a tripod and release the shutter using a cable. You should select the smallest aperture possible on your camera lens, e.g. f16 or f22, and use a slow film such as 50 ISO. The shutter speed can be further extended by the addition of a light reducing filter such as a polarising filter.

Rew - Chris Gannon

Slow shutter speeds

For this technique you can either hold the camera or mount the camera on a tripod. By using shutter speeds slower than those recommended for use with the lens you are using you can create blur from moving objects. If you are using a standard lens try shooting a moving subject at speeds of 1/30, 1/15, 1/8 and 1/4 second. If the camera is in a fixed position on a tripod the background will be sharp but if the camera is panned with the subject the background will produce most of the blur. Try shaking the camera whilst exposing to increase the effect even more.

Fast shutter speeds

For this technique you need to select a wide aperture. Some telephoto and zoom lenses only open up to f4 so you will need to attach a lens that will allow you to open up to f2.8 or wider. Once you have selected the aperture take a light meter reading of your subject. If your indicated shutter speed is only up to 1/250 second then you must do one or both of the following. Either increase the amount of illumination reaching the subject and/or use a faster film.

Multiple exposures

This can be achieved in the following ways:

1. Some modern cameras have a multiple exposure function. If you have a 'classic style' camera that does not have this function double exposure may still be possible. After loading the film but before the exposure, rewind the film using the rewind crank so that there is no slack in the film (do not depress the rewind button at this stage). Place some masking tape or similar over the rewind crank to prevent the film moving. If the background is illuminated or you are shooting outdoors you will need to reduce the exposure, i.e. reduce the aperture by one stop for two exposures and two stops for four exposures. After each exposure and before you wind on to cock the shutter for the next exposure you must depress the film rewind button which is usually on the camera base. If you are shooting against a black background in a studio and the subject is moving away from its original location it will not be necessary to reduce the exposure.

Movement - Alexandra Rycraft

2. Multiple exposures can also be achieved using flash equipment. This is best achieved in a dark studio using a black background. The camera can be mounted on a tripod and the shutter fired using a cable release. The camera's shutter speed dial should be set to B and the cable release kept depressed until the desired number of flashes has been fired. Allow the flash to recharge after each firing. The flashgun will need to be fired manually and you should also ensure you have set the correct aperture on the lens as indicated by your flashgun or flash meter. See the section about working with flash in this study guide for more information.

Zooming

For this technique you need to use a lens that can alter its focal length, i.e. a zoom lens. A slow shutter speed should be selected, e.g. 1/15, 1/8 or 1/4 second. The camera can be mounted on a tripod. The effect of movement is achieved by making the exposure whilst altering the focal length or zooming the lens either in or out. The subject does not need to be moving. Bright highlights and/or bright colours increase the visual effect.

Zooming

Manipulation in the darkroom

A stationary subject or one that has been frozen using a fast shutter speed can be blurred by moving the print slowly during the final seconds of exposure in the darkroom. Movement or time passing can also be achieved by using David Hockney's 'Joiner' technique.

Limitations of semi-automatic

When selecting shutter priority mode on a built-in metering system the photographer must take care not to underexpose images. Excessively fast shutter speeds for the available light may require an aperture greater than that available on the lens.

When selecting aperture priority mode on a built-in metering system the photographer must take care not to overexpose images. Excessively wide apertures to create shallow depth of field in bright light may require shutter speeds faster than that available on the camera.

Activity 5

Create four images that contain a mixture of solid (sharp) and fluid (blur) form.

Flash photography

Flash is the term given for a pulse of very bright light of exceptionally short duration (usually shorter than 1/500 second). When additional light is required to supplement the daylight, flash is the most common source used by photographers. It is difficult to master flash photography because the effects of the flash cannot be seen until the film is processed.
Most flash units available are able to read the reflected light from their own flash during exposure. This feature allows the unit to extinguish or 'quench' the flash by a '**thyristor**' switch when the subject has been sufficiently exposed. When using a unit capable of quenching its flash, subject distance does not have to be accurate as the duration of the flash is altered to suit. This allows the subject distance to vary within a given range without the photographer having to change the aperture set on the camera lens or the flash output. These sophisticated units are described as either '**automatic**' or '**dedicated**'.

Automatic flash

An automatic flash unit uses a photocell mounted on the front of the unit to read the reflected light and operate an on-off switch of the fast acting thyristor type. The metering system works independently of the camera's own metering system. If the flash unit is detached from the camera the photocell must remain pointing at the subject if the exposure is to be accurate.

Dedicated flash

Dedicated flash guns are designed to work with specific cameras. The camera and flash communicate exposure information through additional electrical contacts in the mounting bracket of the gun. The TTL metering system of the camera is used to make the exposure reading instead of the photo cell mounted on the flash unit. In this way the exposure is more precise and allows the photographer the flexibility of using filters without having to alter the settings of the flash.

Setting up a flash unit

~ Check that the film's ISO has been set on the flash and camera.

~ Check that the flash is set to the same focal length as the lens. This may involve adjusting the head of the flash to ensure the correct spread of light.

~ Check that the shutter speed on the camera is set to the correct speed (usually slower than 1/125 second on a 35mm SLR camera using a focal plane shutter).

~ Check that the aperture on the camera lens matches that indicated on the flash unit. On dedicated units you may be required to set the lens aperture to an automatic position or the smallest aperture.

~ Check that the subject is within range of the flash. On dedicated and automatic units the flash will only illuminate the subject correctly if the subject is within the two given distances indicated on the flash unit. If the flash is set incorrectly the subject may be overexposed if too close and underexposed if too far away.

Slow-sync flash

Slow-sync flash is a technique where the freezing effect of the flash is mixed with a long exposure to create an image which is both sharp and blurred at the same time.
Many modern cameras offer slow-sync flash as an option within the flash programme but the effect can be achieved on any camera. The camera can be in semi-automatic or manual exposure mode. A shutter and aperture combination is needed that will result in subject blur and correct exposure of the ambient light and flash.

- Load the camera with 100 ISO film or less.
- Select a slow shutter speed that will allow camera movement-blur (try experimenting with speeds around 1/8 second).
- Take an ambient light meter reading and select the correct aperture.
- Set the flash unit to give a full exposure at the selected aperture.
- Pan or jiggle the camera during the long exposure.

Bowling Alley - Mark Galer

Possible difficulties

Limited choice of apertures - Less expensive automatic flash units can dictate the use of a limited choice of apertures. This can lead to a difficulty in obtaining the suitable exposure. More sophisticated units allow a broader choice of apertures, making the task of matching both exposures much simpler.
Ambient light too bright - A slower film should be used or a lower level of lighting if the photographer is unable to slow the shutter speed down sufficiently to create blur.

Practical assignment

Produce a series of six prints that explore a moving subject by using a variety of techniques and shutter speeds. Your final presentation sheet should demonstrate that you have developed a theme from your initial investigations and that you have chosen appropriate techniques for your subject matter.
A possible title for your set of prints could be:

1. Sport.
2. Dance.
3. Vertigo.
4. The fourth dimension.

Your work should:

a) Make use of different shutter speeds, including some work at or near the slowest and fastest speeds possible with the equipment you are using.
b) Demonstrate an understanding of what is meant by the Decisive Moment.
c) Show that you have considered both lighting and composition in your work.

Presentation of work

Research and all contact prints should be carefully presented on card no larger than A2. Explanatory notes and comments should be made directly onto this sheet or on paper and attached. You should edit your contact sheet including any alterations to the original framing with a chinagraph pencil or indelible marker pen. You should clearly state what you were trying to do with each picture and comment on its success. Contact prints and photographs should be easily referenced to relevant comments using either numbers or letters as a means of identification. You should clearly state how your theme has developed and what you have learnt from your background work and how this has contributed towards the final set of prints.

Resources

Ernst Haas - Bryn Campbell. Collins. London. 1983.
Great Action Photography - Bryn Campbell. Ebury Press. London. 1983.
Henri Cartier-Bresson. Aperture Foundation. New York. 1987.
Motion and Document - Sequence and Time - James Sheldon & Jock Reynolds. 1991.
Photography - London and Upton. Harper Collins College Publishers. New York. 1994.
Photographic Composition - Tom Grill & Mark Scanlon. Fountain Press. 1984.

Books including work by Eadweard Muybridge, Gjon Mili and Harold Edgerton.

Leapfrog - Melanie Sykes

The camera was mounted on a tripod and set to f11 for a 1/8 second exposure.

Sparks - Clair Blenkinsop

The camera's shutter was held open on the B setting in a dark studio whilst a student traced the outline of the model with a sparkler.

Kick - Mark Galer

This slow sync flash effect was created by setting the camera's shutter to 1/15 second, using a 500 watt tungsten light and the flash placed slightly to one side of the subject.

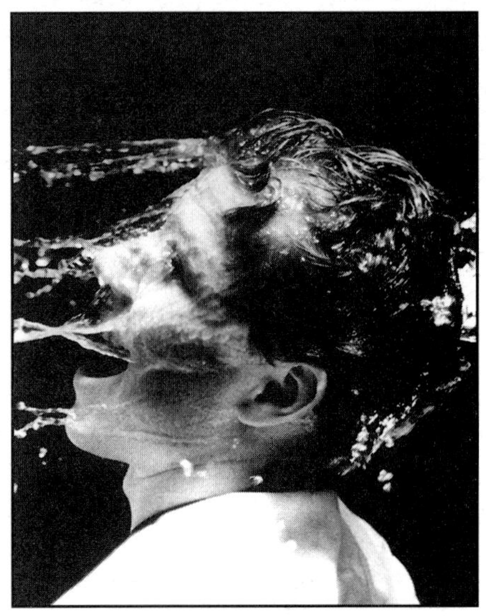

Splash - Claire Ryder

Two 1000 watt halogen lights were used to obtain this fast 1/500 second exposure. Back lighting was essential to illuminate the water against the dark background.

Self Image

Positive and Negative - Faye Gilding

aims

~ To develop an awareness of the links between self image and personal identity.
~ To develop an awareness of symbolism and visual codes of practice.

objectives

~ **Research** - a range of categories that define or divide people and the visual symbols connected with these categories.
~ **Analyse and evaluate** - the effectiveness of the work you are studying and exchange ideas and opinions with other students.
~ **Develop ideas** - produce a study sheet that documents the progress and development of your ideas.
~ **Personal response** - produce and present artwork that explores one or more aspects of your personal character.

Introduction

We are judged as individuals by our appearance, our actions and our status within society. These judgements contribute to our awareness of 'self' and our self image.
The images we choose to represent ourselves help to establish our identity. In photographs we might find evidence of the way we conform, share similarities and fit in with society, e.g. uniforms, fashion, dress code etc. In photographs we might find evidence of what makes us an individual, a unique person within society, e.g. expression, make-up, jewellery, important personal objects, locations and other people.
The images we put forward of ourselves may vary depending on how we would like to be perceived by others. An individual may like to be seen as smart, serious and powerful in a business setting, but casual, friendly and relaxed in a family setting. The images we choose to frame or keep in our albums will tend to portray positive aspects of our self image, e.g. happiness, success and security. This, however, is only one aspect of our 'self image'.

Classifications

Individuals tend to seek out groups that they feel comfortable in. These groups contain individuals who are similar in some respects and together the group forms a collective identity. From this collective identity (team, family group, club or fraternity) we draw a part of our self image that is important to our own identity. Members of large collective groups adopt visual forms of recognition, e.g. items of clothing, body decoration or adornment (scarves, rings, tattoos, haircuts etc.). Along with the visual forms of recognition, the individual is expected to adopt the customs, rituals or ceremonies that serve to unite the group in a common mutual activity, e.g. worship, singing or dancing etc.
The adolescent years and years of early adulthood can be turbulent and explosive for some as they strive either to 'fit in' or 'break out' of some of the categories they were brought up in, or labelled with. As individuals establish new identities conflict may lead to insecurity and anger. The struggle to recognise and define identity is one of life's most difficult tasks.

Activity 1

What follows is a list of categories that serves to define and divide us. Make a personal list of the categories you belong to or have conflict with and visual symbols that are associated with each. Aspirations and expectations should also be listed.

1. Age/generation - This can dictate levels of independence afforded to the individual.
2. Gender - Our sex affects the way we are treated by some individuals.
3. Race - Prejudice and perceived racial superiority can affect hopes and aspirations.
4. Religion - Moral codes and ideologies that serve to guide, unite and divide.
5. Class or caste - A classification that an individual may never escape from.
6. Personality - Extrovert and introvert are classifications of personality.
7. Intelligence - We are often segregated by educational establishments using IQ.
8. Political persuasion - Left wing or right wing. Each carries its own ideology.

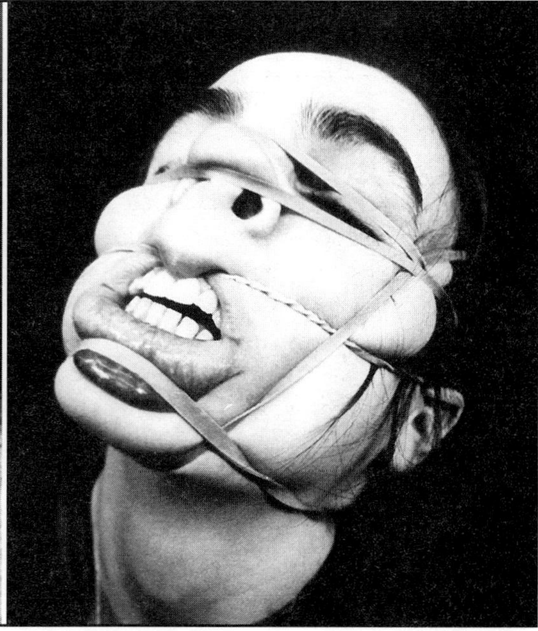

Beauty and the Beast - Catherine Burgess

Images of beauty

The photographs above are of the same student. Each image represents a different aspect of the same character. The image on the left represents how the student would most like to be seen, positive, calm and physically attractive. The image on the right represents how the student would least like to be seen, disfigured, in pain and physically repulsive.

Many people edit images of themselves before they are passed around or placed into a photographic album. People often destroy images that they feel do not portray themselves in a positive way. We are drawn to the image on the right because of its visual impact but we are shocked to learn that the image on the left is of the same student. We do not expect that a single individual has the capacity to be both beautiful and ugly, yet in reality we all feel these reactions to our own image at different times, depending on our state of mind, our physical well being and the manner in which the camera has captured our likeness.

Activity 2

Find images in the media which have been used to represent attractive and unattractive aspects of the human face.

Examine and record carefully the photographic techniques used to accentuate both these qualities, drawing up a list that relates to the images you have found.

List the physical characteristics that we have come to admire in both the male and female face and write 100 words in response to the following questions:

1. Do you believe that media images or public opinion are responsible for the characteristics of beauty becoming universal stereotypes?
2. Do you believe any harm can be done by people admiring media images of glamourous models?

Fashion Statement - Thomas Scicluna

Images of conformity

The photomontage above shows the same figure dressed in many strange ways. Each figure is accompanied by a bar code as seen attached to the products we buy in the supermarket. The orderly lines of figures contrast with the main figure in the foreground who is wearing his underpants outside his trousers and who stares straight at the viewer in an unnerving manner.

The student is making the statement that fashion manipulates the young. Fashion rather than being an individual attempt at personal expression is a commercial way of pressurising people to conform (no matter how silly the conformity looks). The main figure looks like he has just recognised this fact and is not amused.

Boxes - Housewife - Julia McBride

The photomontage above expresses the student's awareness of social pressures to conform to the image and behaviour of a typical housewife. The student is seen to be breaking out of the 'box' in which she feels she may be placed at some future date.

Activity 3

Consider some of the social pressures that you think may shape your behaviour and personal image. List the images most commonly associated with the categories or 'boxes' you have already listed in Activity 1.

How have you responded to social pressures to conform by adopting an appearance that relates to the categories that you feel you have been placed in or have chosen?

Practical assignment

Using photography and other media put together a composite image or series of images that communicate how you see yourself in harmony or conflict within the social structure of which you are a part. The assignment may put forward one aspect or several concerning your self image and it is envisaged that you will explore not just the positive ones.
Your design sheets or sketch book should show evidence of the development of a variety of ideas or approaches to this assignment and together with the finished piece of work they will form an integral part of the assessment material.
Possible starting points for practical work concerning self image:

1. Reflections:
~ a window into the soul?

2. Different aspects or sides to our character could be illustrated through conflicting, contrasting or changing images:

~ Negative and positive, strong and weak, child and adult etc.

~ Different aspects of our character viewed simultaneously or consecutively.

~ Triptychs, installations, constructions, e.g. cubes, cylinders, mobiles etc.

~ Our dreams and aspirations versus the reality we inhabit.

~ A recognition of the sources of individual pressures within society to conform or adjust our behaviour and appearance.

~ Stereotypes - racial, gender etc.

Resources

Media images representing age, class, gender and race.

The family photograph album.

Personal collections of photographs.

The photographic work of Jo Spence.

The photographic work of Cindy Sherman.

Façades - Daniel Shallcross

Daniel has created this image by breaking a mirror into many pieces on the floor and photographing the reflection of a single individual. Due to the shallow depth of field the actual fragments of the mirror cannot be seen. Daniel was working on the idea of an individual possessing more than one identity and how the image may be a mask to the true identity.

Reflection - Claire Ryder

Claire has used a highly reflective piece of card which has been bent to produce this dramatic effect. The model and their reflection can be viewed at the same time but the reflection is a distorted view of reality. Claire has produced an image which explores how our self image can be distorted by our levels of confidence and by the recording methods used.

Gallery

Matthew Houghton

Matthew Theobold

Photomontage

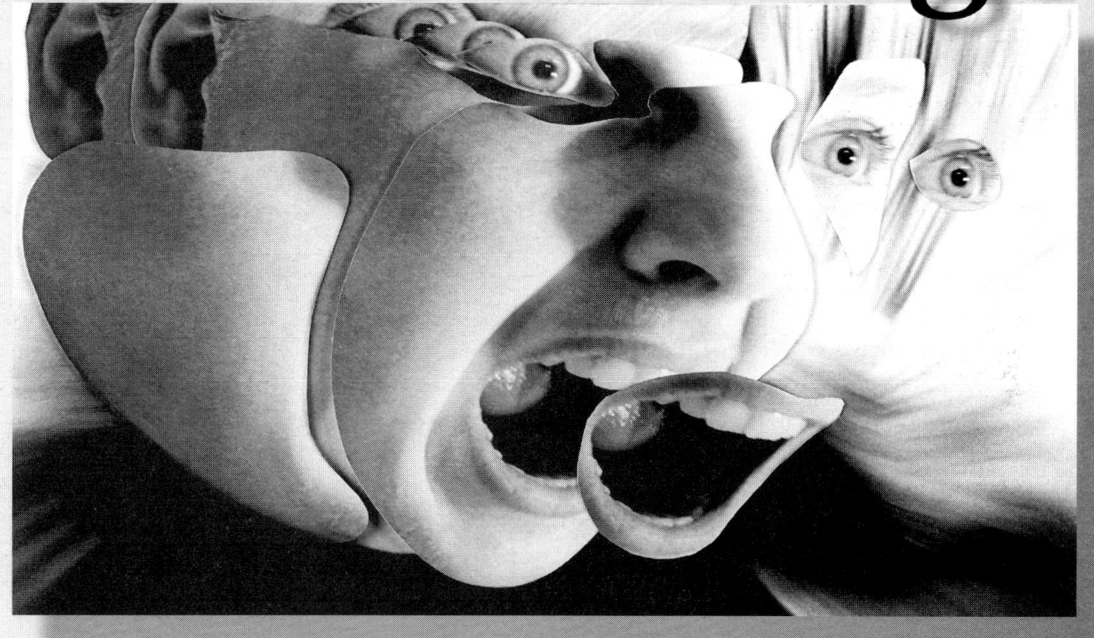

Agoraphobia - Chris Gannon

aims

- ~ To develop an awareness of how visual elements interact with each other and with text to convey specific messages.
- ~ To develop an awareness of the techniques employed by individuals in the media in order to manipulate photographic material.

objectives

- ~ **Research** - a range of assembled photographic images from both commercial and non-commercial sources.
- ~ **Analyse and evaluate** - the effectiveness of the work you are studying and exchange ideas and opinions with other students.
- ~ **Develop ideas** - produce a study sheet that documents the progress and development of your ideas.
- ~ **Personal response** - produce and present artwork to demonstrate how assembled photomontages can communicate a specific message.

Introduction

When we look at a photograph we read the images as if they were words to see what information they contain. If the photographic image is accompanied by words they can influence our decision as to what the photograph is about or even alter the meaning entirely. The photographer can manipulate the message to some extent by using purely photographic techniques such as framing, cropping, differential focusing etc. This controls which information we can or cannot see. Some artists and photographers have chosen to increase the extent to which they can manipulate the information contained in a photograph by resorting to the use of paint, chemicals, knives, scissors and electronic means to alter what we see. The final result ceases to be a photograph and becomes a '**photomontage**'.

Definition

A photomontage is an image that has been assembled from different photographs or from a single photograph that has been altered. By adding or removing information in the form of words or images the final meaning is altered. The resulting photomontage may be artistic, commercial, religious or political.

Destruction - Darren Ware

History

Photomontage is almost as old as photography itself. During the 1850s, several photographers and artists attempted to use photography to emulate the idealised scenes and classical composition that was popular in the Pre-Raphaelite painting of the period. Photographs during this time were restricted by the size of the glass plate used as the negative in the camera. Photographic enlargers were not common, so the images had to be contact printed onto the photographic paper. By using many negatives to create a picture the photographer could increase the size of the final image. This also released the photographer from the need to photograph complex sets using many models. The technique of 'combination printing' allowed the photographer to photograph each model individually and then print them on a single piece of paper masking all the areas of the negative that were not needed.

Historical examples

A famous combination print of the Victorian era is called 'Two Ways of Life' by Oscar Gustave Rejlander, produced in 1857. As many as 30 negatives were used to construct this image. The picture shows the moral choice between good and evil, honest work and sin. Another photographer who worked with this technique was the artist Henry Peach-Robinson. Robinson first sketched his ideas for the final composition on a piece of paper and then fitted in pieces of the photographic puzzle using different negatives. One of his most famous images is the piece titled 'Fading Away' which was constructed in 1859 from five negatives. The image depicts a young woman dying whilst surrounded by her family.

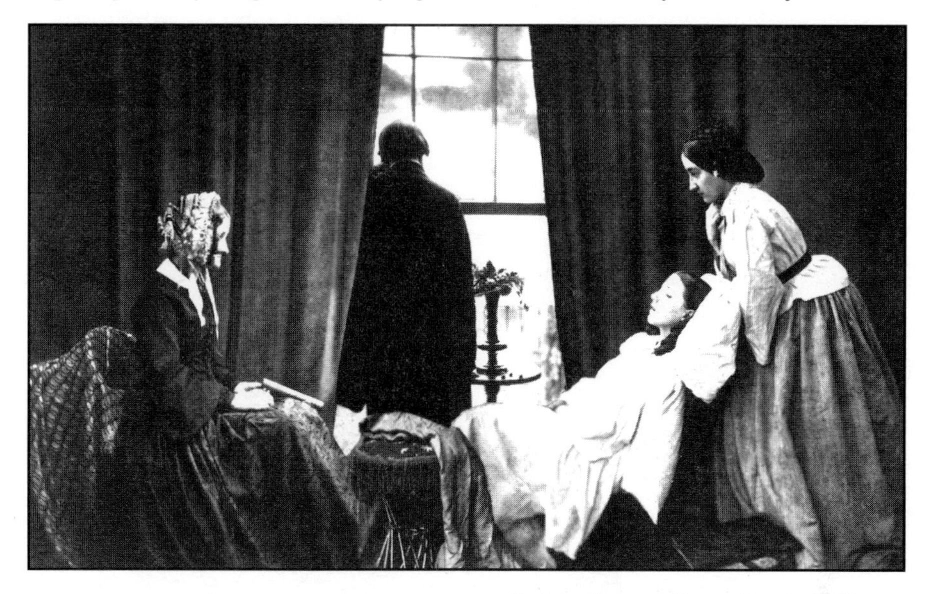

Fading Away - Henry Peach-Robinson

Historical developments

Both Peach-Robinson and Rejlander were heavily criticized by the fine art world for their work in this field but the technique was popular with the public and did survive. Commercial photographic studios in the 1860s started making their own composite photographs. They were made by cutting up many photographs, usually portraits of the famous or of beautiful women, and gluing them onto a board. This was then re-photographed and the resulting prints sold as souvenirs or given away as promotional material.

Activity 1

Briefly discuss why you think painters of the Victorian period might have criticised these early photomontages and why the pictures were popular with the general public.
This technique of piecing together separate images to create one picture is again very popular with both artists and the media.
What two reasons can you think of for this revival of an old technique?

Political photomontage

The technique of photomontage was not widely used again until the Cubists and the Dadaists in the 20th century experimented with introducing sections of photographs and other printed material into their paintings. A Dadaist by the name of John Heartfield further developed this use of photography and is now commonly attributed as the founder of political photomontage.

John Heartfield was influenced by Dadaism and the newly emerging socialism in the Soviet Union. Sergei Tretyakov, a Russian born writer, showed Heartfield how a collection of facts carefully edited could convey a significant message. John Heartfield was already deeply involved with politics and he now saw how the graphic use of assembled photographs, combined with the mass production printing techniques of the time, could reach a wide audience with his political messages.

Defended to Death - Peter Kennard

The techniques that John Heartfield developed in the 1920s and 1930s are still used by contemporary artists such as Klaus Staeck and Peter Kennard. In the 1980s Peter Kennard used photomontage to draw attention to the escalating arms race, his most famous works being associated with the organisation CND (the Campaign for Nuclear Disarmament). Both artists have found photomontage an ideal medium to communicate social and political injustice to the public in an immediate and effective way. Through photomontage the public are asked to question the media images they see everyday, thus raising their visual literacy.

Techniques of political photomontage

1. **Text:** Text included with most photomontages sets out to reinforce the message of the photographer or artist. The words can remove any possible ambiguity that the viewer may find, draw the attention to the conflict in the images or contradict the images entirely thus establishing a satirical approach to the work.

2. **Recognising the familiar:** The viewer is meant to recognise familiar images, paintings, photographs, advertising campaigns etc. that the photomontage is based upon. The viewer is drawn to the differences from the original and the new meanings supplied by the changed information. This technique is often used in contemporary photomontage works. The technique is to make the familiar unfamiliar, e.g. 'Cruise Missiles' Peter Kennard, 'Mona Lisa' Klaus Staeck.

3. **Contradiction:** In many photomontages we see images that contradict each other. Introduced parts of the photomontage may be inconsistent with what we would normally expect to see happening in the image thus questioning the original point of view. The contradiction can also be between what we see is happening and the text. Our curiosity to establish a coherent meaning in both cases is raised.

4. **Contrast:** To gain our attention individual components of the photomontage may be in sharp contrast to each other, e.g. wealth and extreme poverty, tranquillity and violence, happiness and sadness, industrialisation and the countryside, filth and cleanliness etc.

5. **Seeing through the lies:** Many photomontages invite the viewer to look behind the surface or see through something to gain greater insight into the truth. Windows that open out to a different view, X-ray photographs that reveal a contradiction are all popular techniques.

6. **Exaggeration of scale:** By altering the scale of components of the photomontage the artist can exaggerate a point, e.g. a photomontage entitled 'Big business' may include giant, cigar smoking, industrialists crushing smaller individuals under their feet.

7. **Figures of speech:** A very popular technique of the photomontage artist is to visualise figures of speech, e.g. puppet on a string, playing with fire, house of cards etc. The artist may play upon the viewer's acceptance of these as truth or use them as a contradiction in terms, e.g. 'The camera never lies'.

Activity 2

Find two examples of political photomontages that are either from a historical or contemporary source. Discuss in what context they have been produced and how effective you think they communicate their intended message.
Discuss the techniques that have been used to assemble the examples you have chosen and offer alternative ways that the artist could have put over the same message.

Photomontage in the media

Advertisers and the press have embraced the new technology, which makes image manipulation easy, with open arms. It is fast, versatile and undetectable. Advertising images are nearly always manipulated in subtle ways to enhance colours, remove blemishes or make minor alterations. We are now frequently seeing blatant and extensive manipulation to create eye-catching special effects. We do not get unduly agitated or concerned over these images because advertising images have always occupied a world that is not entirely 'real'.

To manipulate or not to manipulate

We start to move into more questionable territory when images we take to be from real life, perhaps of real people, are altered without our knowledge. The top fashion models we see on the covers of magazines like *Vogue* and *Vanity Fair* for instance, we believe to be real people, especially when they have been taken out of the advertising context. We expect models to have near perfect features and complexions, but it would come as a shock to many sections of the public to realise that these images are frequently altered as they are not quite perfect enough for the picture editors of the magazines. The editors do not distinguish between the images that appear on the cover of the magazine and the images that are sent to them in the form of advertisements. Minor blemishes to the model's near perfect skin are removed, the colour of their lipstick or even their eyes can be changed to suit the lettering or the colour of their clothes. Many women aspire to these models and their looks, and yet these people don't exist. It is important to remember that a publication is a product, whether it is a fashion magazine or a daily newspaper, and as such editors may be more interested in sales than in truth.

Most editors are at present using their own moral codes as to when, where and to what extent they will allow manipulation of the photographic image. Most editors see no harm in stretching pictures slightly so they can include text over the image or removing unsightly inclusions. What each editor will and will not allow can vary enormously. Occasionally the desire for an image to complete or complement a story is very strong and the editor is tempted to overstep their self-imposed limits. A fabricated image can change the meaning of a story entirely or greatly alter the information it contains. It is possible to construct a visual communication where none may otherwise have existed.

Activity 3

As editors exercise their ever increasing power over information control, what limits would you impose on them as to the extent to which they can manipulate the photographic image? Devise a series of guidelines that will control the release of images that have been constructed for media use so that the public are aware as to the extent of the manipulation.

Activity 4

You have been commissioned to produce a series of posters that aims to increase awareness about the effects of image manipulation. The posters are intended as part of a study pack for A level Media Studies students. The pack will include a booklet of photographs and supporting text about this important subject.
The posters should illustrate the following statement:

A photograph, whether it appears in an advertisement, a newspaper or in a family album, is often regarded as an accurate and truthful record of real life. Sayings such as 'seeing is believing' and 'the camera never lies' reinforce these beliefs. The information we see in photographs, however, is often carefully selected, so what we believe we're seeing is usually what somebody would like us to see.
By changing the information we can change the message.

The publishing company of the study pack envisages a series of photomontages or image manipulations that will cover the implications of image manipulation in all areas of the printed media.
Produce **ONE** finished poster in colour or monochrome. You may use a mixture of black and white with coloured text. With the finished poster you should submit a set of layouts and your preliminary sketches. The work that you submit should include the following:

1. The final poster should be supported with at least four different design layouts at a reduced scale. These layouts should demonstrate a variety of creative approaches to this assignment and include the text that would accompany the image. You should have considered designs for posters dealing with both press and advertising manipulation.
2. The final choice of poster should be supported with a 200 word 'rationale' that explains how the poster will raise general awareness to the issues of image manipulation. The rationale should also include the techniques you have used to communicate this message.
3. Text included in the final design could be one of the sayings 'seeing is believing' or 'the camera never lies' which should conflict with the image that you produce or it could be some copy of your own that reinforces the message conveyed by your image.
4. The size of the final poster should either be A2 or A3 and include the chosen text in a size that can be easily read from 3 metres.

Photomontage in art

Around the time that John Heartfield was developing the language of political photomontage, another artist, Laszlo Moholy-Nagy, was using photography in conjunction with drawn geometric shapes. The effect was to cause a visual conflict between the print viewed as a factual record of three-dimensional reality and the print as a two-dimensional surface pattern. This important work ensured that many contemporary artists would continue to use the camera as a creative tool. Some have found 'straight photography' limited as a medium for visualising their own personal ideas or abstract concepts within a single image. Image manipulation and photomontage are two ways that photography can be used by the artist to communicate in more complex ways.

Surrealism

Surrealism followed Dadaism and photography was found to be an ideal tool and so became directly involved in this movement as it had been in Dadaism.
Surrealist images are constructed from the imagination rather than from reality. Many of the works by surrealist painters such as Salvador Dali and Réne Magritte appear as fantastic dream sequences where familiar objects are placed in unusual settings. Parts of the image may be distorted in scale or shape and the impossible is often visualised. Inspired by the work of these surrealists Angus McBean, in the 1930s, produced a whole series of photomontages of actors and actresses. The photographs are a strange mixture of fantasy and portraiture where the character is often surrounded by props from their current play.

Audrey Hepburn - Angus McBean

Joiners

David Hockney has used photomontage in many of his works. Individual photographs have been placed together to create a larger photographic collage which Hockney has termed a 'Joiner'. The number of prints in each Joiner may vary depending on how much information Hockney feels he needs to explore in the subject matter. In some of his Joiners he has photographed more than one side of an object or person and then placed the individual images together to create a single picture. This technique was used by cubist painters such as Georges Braque and Picasso who show many facets of objects on a single canvas. Hockney has taken this one stage further and presented the subject at different moments in time. The photographic instant has been lengthened in some of Hockney's Joiners to become a study of a 'happening', different views from different times during the same event. A picture of a card game is no longer a record of that game at one particular moment but rather a study of the whole game.

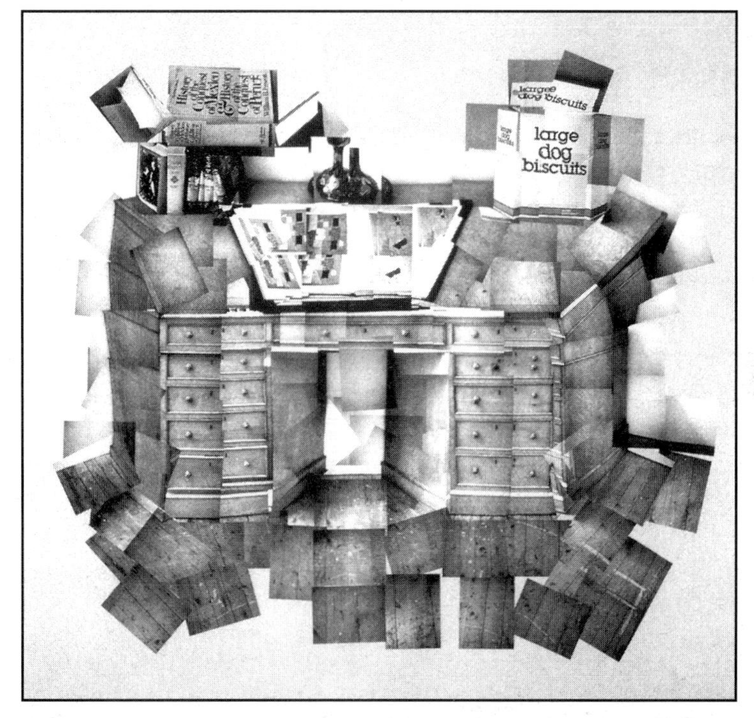

The Desk - David Hockney

Activity 5

Find two examples of photomontages that are either from a commercial source or from a fine art background. Discuss in what context they have been produced and what techniques they share with political photomontages.

What messages, if any, are communicated through these photomontages and how effective do you think they are? Consider different ways that each image could have been tackled by the artist and come up with an idea for 'another in the series'.

Practical assignment

Produce one A2 photomontage or a series of smaller montages in response to one of the following titles:

a) **Phobia** - an abnormal or morbid fear of something
b) **Metamorphosis** - a change of form, character or conditions
c) **Decay** - decompose
d) **Dream world**
e) **Strange but true**
f) **Fast food**
g) **Fashion**

Additional information

Your work should include:

1. A detailed description of how you have photographed each individual element of the montage and the sources for 'found' images that you may have had to use (images taken from the printed media).
2. A description of how the final montage has been assembled.
3. A research sheet including your own contact sheets and ideas.

Resources

David Hockney Photographs - Cameraworks. Thames and Hudson. London. 1984.
Images for the End of the Century - Peter Kennard. Journeyman Press. 1990.
In Our Own Image - Fred Ritchin. Aperture Foundation. New York. 1990.
Jerry N. Uelsmann, twenty five years: a retrospective. Little, Brown. Boston. 1982.
Photomontage - a Political Weapon - Evans and Gohl. Gordon Fraser. London. 1986.
Photomontage Today - Peter Kennard (35 minute video). Arts Council of Great Britain.
Photomontages of the Nazi period - John Heartfield. Universe Books. New York. 1977.

Books including work by the following artists and photographers:

Historical Photomontage - Oscar Gustave Rejlander & Henry Peach Robinson.
Political Photomontage - John Heartfield, Klaus Staeck & Peter Kennard.
Surrealism - Réne Magritte, Salvador Dali, Laszlo Moholy-Nagy & Angus McBean.
Contemporary Artists - David Hockney, Jerry Uelsmann.

Eating Cars - Zara Cronin

Space & Pavement - Darren Ware

Gallery

Phobia - Lizette Bell

Distortion

Distortion - Ashley Dagg-Heston

aims

- To extend personal creativity through improved technique.
- To explore the limitations of photographic materials and equipment through experimentation and exploration.

objectives

- **Research** - locate and investigate various techniques of photographic distortion as used by historical and contemporary photographers.
- **Analyse and evaluate critically** - exchange ideas and opinions with others.
- **Develop personal ideas** - produce a study sheet that documents the progress and development of personal ideas in response to your research.
- **Personal response** - produce and present images that explore a range of techniques and processes to communicate a personal theme or idea.

Introduction

All photographs distort reality to a lesser or greater degree. The very act of converting three-dimensional reality into a two-dimensional print is a distortion in itself. The simplest of photographic techniques used by the photographer and editor can manipulate and distort the information that the photograph puts forward to the viewer. This can be achieved in the following ways:

- the act of framing composes the visual elements within the frame and decides which of the visual elements we can or cannot see in the final photograph;
- the viewer can be led through the print in a systematic way through the use of line, tone and colour;
- the choice of lighting can effect the mood of the image, to reveal or disguise form and texture;
- use of shallow depth of field called 'differential focusing' can guide the viewer to individual elements within the frame;
- movement can be explored or the illusion of movement created through the controlled use of shutter speed and flash lighting;
- visual elements can be removed or added after the photograph has been taken to change the meaning of the image;
- the use of text can be used to clarify or contradict meaning.

The limits of distortion

Photographic techniques control the way we read information from photographs in many ways, yet most people accept the accuracy of the photographic medium in recording the subject matter in front of the lens. This study guide is designed to give you the opportunity to discover the limits of distorting this information using both the equipment and the materials commonly used in the photographic process.

Activities

a) Experiment with five of the techniques outlined in this study guide.
Use subject matter you feel is appropriate to the visual effect.
Record the procedures you have used accurately so that the process can be perfected at a second or third attempt.

b) Research both art and media sources for examples of photographic distortion and make notes as to how you think this effect may have been achieved.

Note. The final practical assignment requires that you work within a theme (as with previous assignments). It is advisable that you consider this theme whilst working through the activities section of this study guide. This will ensure that your research work is seen as appropriate.

Distortion using the camera

Lens distortion

Telephoto or long lens - A telephoto lens compresses, condenses and flattens three-dimensional space. Subject matter appears to be closer together. This effect is most noticeable when the lens has exceeded the focal length of 135mm.
Very long lenses are expensive but extreme effects can be explored by attaching a teleconverter to a shorter focal length lens. A focal length of 400mm or greater may be achieved in this way without paying large sums of money. The problems encountered by using this technique are maximum apertures that may be no better than f8. A fast film will be necessary in these situations.

Wide-angle lens - A wide-angle lens exaggerates distances and scale is distorted. Subjects close to the lens looks larger in proportion to their surroundings. Subject matter in the distance looks much further away. The overall effect is one of 'steep perspective'.
A wide-angle lens with a focal length of 28mm or shorter is recommended to explore this technique. The closer you move to the subject the greater the distortion.

Zooming - For this technique you need to use a lens that can alter its focal length, i.e. a zoom lens. The camera can be mounted on a tripod and a slow shutter speed selected, e.g. 1/15, 1/8 or 1/4 second. The effect is achieved by altering the focal length during the exposure. The subject does not need to move for the effect to work (see '**Time**').

Depth of field - Very short depth of field is achieved by using one or more of the following techniques: a) long focal length lens, b) wide aperture, c) moving closer.
Wide depth of field is achieved by using a combination of one or more of the following techniques: a) short focal length lens, b) small aperture, c) moving further from subject.

Refraction and reflection

Refraction - This is the action of light being bent or deflected as it passes through different media such as glass and water. Look at the swimmer photographed by André Kertész. Try photographing through patterned or textured glass, special filters, clear filters with Vaseline smeared on them, water etc. in front of the lens.

Reflection - When light is reflected off smooth surfaces which are curved we get a view of the image distorted. Look at the nudes produced by André Kertész.
The mixing of reflections with the view through plane glass can also produce interesting effects, as can mirror images introduced into the picture. Possible sources for such images include chrome items, reflective foil, shiny black cars etc.

Lighting and film

Artificial light - Try using unusual lighting conditions such as car headlights or portable video lights to illuminate your subject. Experiment mixing tungsten and flash or by placing opposite coloured filters over the camera lens and a flashgun. Try experimenting with coloured filters using black and white film (a red filter turns a blue sky virtually black).

Infrared film - Infrared is a part of the light spectrum that is invisible to the human eye but can be recorded on special film. Skies appear dark and green vegetation appears to glow. This film needs to be used in conjunction with a deep red filter and must only be handled in complete darkness.

Lith film - This is a very high contrast film which can be loaded directly into the camera and exposed at 6 ISO. The film can be developed in special lith developer so that hardly any grey tones remain.

Point Blank - Philip Budd

Depth of field

The photograph above draws the viewer's attention to the barrel of the gun. The student has used a mixture of telephoto lens, close proximity to the subject and a wide lens aperture to create this photograph. This is a technique known as differential focusing and is an example of pushing the equipment to its limits (see '**Light**').

Distortion in the darkroom

The negative

Little can be done to gain any worthwhile effects during the processing of black and white negatives. The following can be attempted but the effects are unpredictable.

a) **Reticulation -** This is a process designed to damage the grain structure of the negative by subjecting the film to extreme temperature changes during processing. Replacing the stop bath with very hot water and then ice cold water before fixing the film can achieve reticulation. Most modern films are very resistant to this type of abuse or treatment.

b) **Solarisation** - This is a process of fogging the film to white light during the development stage of the negative. The resulting negative, if the process is carried out correctly, will appear half positive and half negative. The effect is more commonly achieved in the printing stages due to the unpredictable nature of the process and is often called pseudo-solarisation or sabatier. If you are going to attempt it using film, try using a high contrast film such as 'lith' film in the camera and concentrate on subjects with bold patterns and that have been lit with a hard directional light source.
The flash of light required to fog the film during the development stage must be of the correct intensity. Try experimenting with short pieces of film loaded onto a film spiral. Remove the developing tank lid in the darkroom (with the red safelight switched off) shortly before half the developing time has elapsed. Place the tank on the baseboard of an enlarger underneath the enlarging lens. Expose the film whilst it is still submerged in the developer to 1 second of light with the enlarger lens on a small aperture. The fogging must occur when the film is only half way through the development stage. Once the film is fogged no further agitation should be carried out until the development stage has been completed. If the resulting negative is too dark after it has been fixed reduce the aperture further or move the head of the enlarger higher on the column. Increase the exposure if no effect can be seen.

c) **Extreme grain** - To achieve very large grain try using hot developer for a reduced amount of time and/or the use of Kodak recording film.

d) **Cross processing** - This is a technique whereby a colour transparency film is processed in chemicals designed to process colour negative material (C41 process). The result is an interesting shift in colours.

e) **Distortion of the film emulsion** - It is possible to mark or manipulate the film emulsion in a variety of ways. Any manipulation of the negative surface is usually permanent so it is recommended that several identical shots of the same subject are taken so that individual techniques can be perfected without the loss of a good shot. The film emulsion can be scratched using a sharp instrument such as a compass or marked with pens and paint. Special opaque paint called 'photo-opaque' can be used on film surfaces. This will mask any areas that you do not wish to print. Negatives are constructed from a celluloid layer which can be distorted with chemicals or heat. Black and white emulsion layers are very resistant when dry but colour transparency emulsion layers will bubble when heat is applied.

Tone elimination

Tone elimination reduces the number of tones or shades of grey in an image. This technique is used in screen printing where only one tone or colour can be printed at any one time.

Normal print

a) Preparing the materials - You will need lith film (either sheet film or 35mm) and lith developer. A contact printing frame will also be necessary if you are using 35mm film (ensure that the glass is clean). The film can be handled in darkrooms that use red safelights.

b) Making a test strip - Enlarge the original negative onto the sheet film or contact print the strip of negatives onto lith film, emulsion to emulsion (the emulsion side of lith film is lighter in tone). Make a series of timed exposures as if you were making a test strip for a print.

A two tone print

c) Processing - Process the lith film in a tray with lith developer at 20½C for 2-3 minutes (lith developer deteriorates rapidly once mixed). Wash and fix until clear.

d) Making a positive - Choose the best exposure and repeat the process. Wash and dry the positive well. If you require more than one tone you will need to make positives of varying exposures. The third image opposite was made using three different exposures.

e) Making a new negative - Repeat the previous process using the positives instead of the original. Wash and dry the negatives well.

A four tone print

f) Printing the tone elimination - Retouch any dust marks on the lith negatives using photo-opaque, a fine paint brush and a light box prior to printing. If you are using only one lith negative to make your print follow your usual printing procedure. If you are printing from several lith negatives you will need to register each negative by using registration marks. Mark important parts of the image with a pen on a piece of paper inserted into the printing easel. Register each negative to these marks before reinserting the photographic paper. Each exposure will be of the same length, the combined exposures creating the darkest tone in the shadow areas.

The print

a) Negative prints - These can be produced from any dry print and are often produced from solarised prints (see the work of Man Ray). First make sure that the light from the enlarger lens covers the baseboard and set the enlarger lens to the brightest aperture. Place a fresh sheet of unexposed printing paper, emulsion side up, on the base-board. Place the original print face down onto this unexposed printing paper. Place a clean sheet of glass onto the two sheets of paper to ensure close contact or use a purpose built contact printing frame. Make a test strip to find out the correct exposure. Exposures will be longer than if you had used a negative because the light has to pass through the top print to expose the paper underneath. Process the test strip and the resulting negative print as normal.

b) Solarisation - This is a technique which gives the print the appearance of being both negative and positive at the same time. It is achieved by subjecting the print to a brief second exposure during the development stage. The lightest tones of the print are affected most by this second exposure thus altering the normal tonal values of the print. A white line or halo appears between the first and second exposures. See '**Making a solarised print**'.

c) Selective developing and fixing - The processing chemicals can be applied to some of the areas of the printing paper leaving others to fog or remain undeveloped. Many imaginative ways can be found to apply the chemicals or prevent their contact with the print surface. These include spraying, painting or dripping the chemicals and by coating areas of the print with hand cream, Vaseline or masking fluid which can be removed before fixing.

d) Movement of printing paper - The printing paper can be twisted, bent or dragged on the enlarger baseboard for a varying length of the exposure to distort the image or give the illusion of movement.

e) Toning - Many chemicals are available to tone photographs that have already been processed. These include sepia (a warm brown tone), blue and copper toner. The prints should be soaked for half an hour in water. Blue toning a print will darken it, so it may be advisable to start with a print a little lighter than usual.

f) Colouring - Paints, inks and dyes can be applied directly to the finished print. Gloss and resin coated papers are more difficult to work with than matt and fibre-based papers and will generally only accept acrylic paints.

g) Sandwiching negatives - This is achieved by placing two negatives together in the enlarger carrier (emulsion to emulsion). Experiment with using one negative of a silhouette and one negative of a surface texture or pattern. If exposures are too long try using negatives that are slightly underexposed.

h) Multiple exposures - This can be achieved by exposing different negatives onto the same piece of printing paper or turning the paper around and re-exposing the paper using the same negative. If a subject has been photographed against a bright background this will produce dark areas on the negative. When the negative is printed areas of the print will be unexposed, allowing a second exposure to be made into this area. This technique will require some planning to be effective.

Making a solarised print

a) Selecting an image - This should be a bold image with a strong pattern. The image could contain blocks of differing tone or strong graphic lines.

b) Make a test strip - Using grade 4 or 5 makes a high contrast test strip. If the negative will not give you a high contrast print choose another negative.

c) Make a print - Choose the best exposure from the test strip and reduce the exposure by 25-30%. Make 2 or 3 prints but do not process them. Store them in a box or black plastic bag.

d) Preparing for the second exposure - Remove the negative from the enlarger. Make sure that the light from the enlarger lens covers the baseboard of the enlarger either by raising the enlarger head or altering the focus. Cover the baseboard with paper towels and place a developing tray half full of water onto these. The light from the enlarger must cover all of the tray. Switch off the enlarger.

e) Make a second test strip - The aim of making this second test strip (without the negative) is to find a middle grey tone. This tone will indicate the time required for the second exposure. Set the enlarger timer.

f) Half process the print - Process one print that you made earlier but remove it from the developer when it has received only half the developing time. The image should only be half the correct density. Slide the print gently into the tray of water and let the surface settle.

g) Second exposure - Switch on the enlarger to give the print its second exposure.

h) Final processing - Return the print to the developer and process normally. Fix and wash as usual.

i) View the print - View in daylight conditions and alter the time of the second exposure if needed.

Normal print

Solarised print

Negative of solarised print

Computer manipulation

The images opposite demonstrate how copying and pasting can be used to manipulate the original image. The first image shows a man with a broken nose and one half of his face in shadow. It was decided to duplicate the left-hand side of the photograph to produce the manipulated image at the bottom of the page.

The original image

Cut, copy and paste

The electronic imaging software has several tools which allow the operator to cut out or copy sections of an image and paste them in a different place or in a different image entirely.

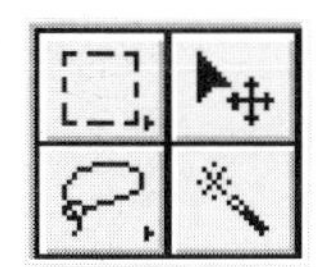

The tools above are:

Top left - Rectangular marquee tool
Top right - Move tool
Bottom left - Lasso
Bottom right - Magic wand

The selection is moved

A selection using the marquee tool is made by first clicking on the tool and then by clicking on the image and dragging the mouse until the area of image that you require is indicated by a dotted line.
The lasso is used to draw around a section of the image whilst the magic wand is placed on a specific area of the image and clicked. The magic wand will select all the pixels that are either the same or are similar to the ones where the magic wand was placed.
The selection is then copied and moved, leaving the original image intact.
The selection is flipped horizontally so that we then view a mirror image of itself. When the selection is correctly positioned it is deselected. Any areas that don't quite match can then be retouched to produce a manipulation that is very difficult to detect (see '**Digital manipulation**').

The selection is reversed and deselected

Practical assignment

Choose **one** assignment only. All work should be supported with research and background work. This should include the work carried out in the 'activities' section of this study guide plus evidence of how your ideas have developed for your chosen theme.

1. A publishing company has commissioned you to produce a set of photographs that will advertise their forthcoming book titled *Beyond Reason* one of which will be used on the dust jacket of the book.
The book is about psychic phenomena and it is envisaged that the photographs will use special effects to set the mood for the articles in the book.

2. You have been commissioned by a successful photographic studio to establish and promote a specialised area of photography. The area is to be known as '**The Dirty Tricks Department**'. This will be a mixture of specialised photographic effects together with original creative thinking.
The promotion is to be a leaflet containing a number of images and you may wish to include the title as part of the presentation.

3. Choose one of the following themes from which to work. Think very carefully about the many ways you can approach your chosen theme. Do not choose the most obvious definition of the word unless you have considered alternatives.

Sensation
Force
Transitions
Rituals
Faith
Masks
Dreams

4. Produce a set of promotional photographs that would be suitable for the pressure group Amnesty International. The images you produce should promote a particular aspect of their work that you feel strongly about.

5. Express in photographs one or more of the following emotions. You may consider using props and/or text to convey the emotion effectively.

Anger
Fear
Love
Hate
Joy

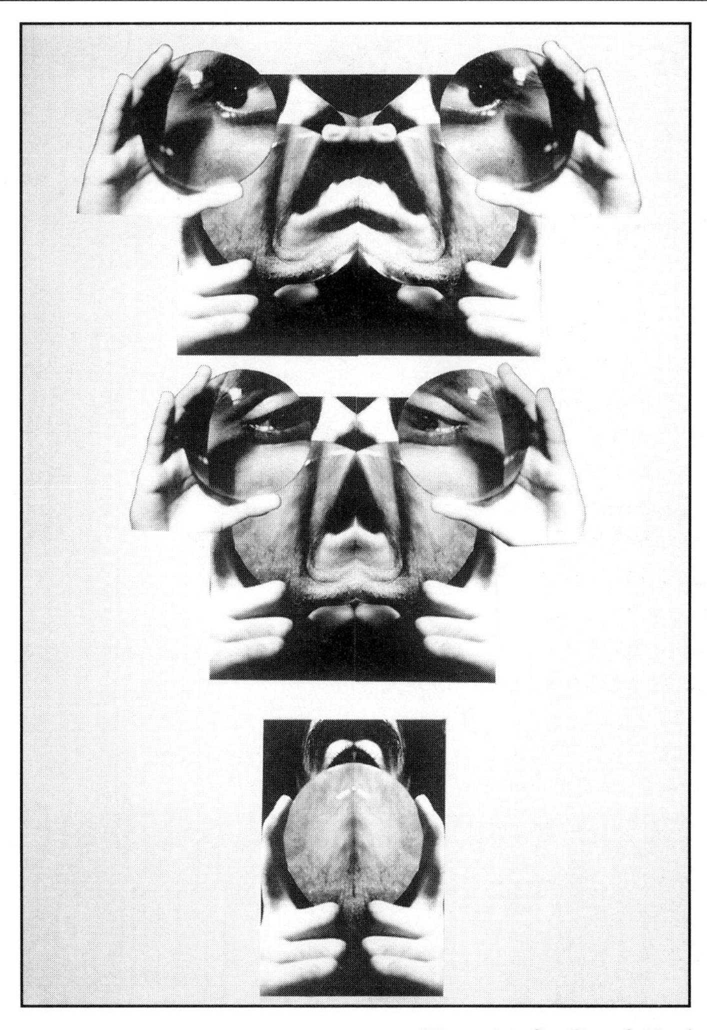

Distortion 2 - Gareth Neal

The student has created a sequence of distorted images. A magnifying lens has been used to distort sections of the human face whilst taking the shots. These have been further distorted by using montage techniques to piece them together with mirror images of themselves. The sequence shows the original image at the top disappearing into itself.

Resources

André Kertész - Nudes .... distorted reflections
Bill Brandt - Nudes 1945-1980 .... use of wide angle lens distortion
Classroom Photography - Ilford (London) .... creative use of materials
Creative Darkroom Techniques .... Kodak
Ernst Haas .... use of slow shutter speeds
Man Ray .... solarisation, negative prints, multiple exposures and photograms
The Print, Time Life Publications .... images created in the darkroom
The Workbook of Darkroom Techniques ....John Hedgecoe/Mitchell Beazley

Faces - Paul Heath

This print is a double exposure which has then been solarised whilst processing the print.

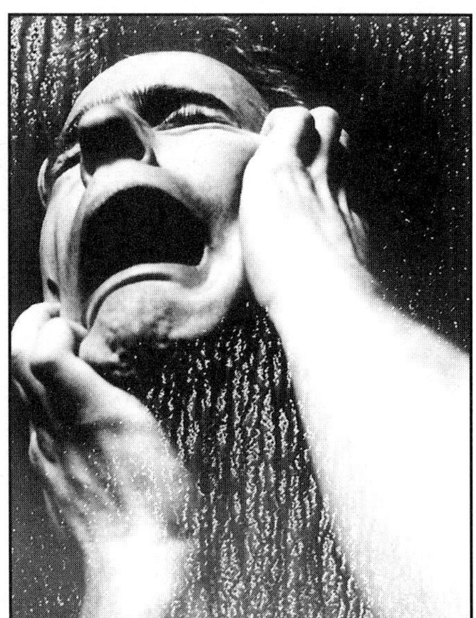

Scream - Lynsey Berry

This image was distorted by spraying the developer from a plant sprayer whilst the developing tray was tilted.

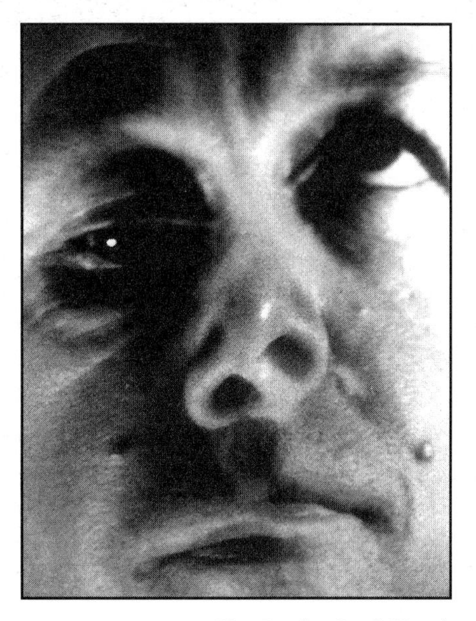

The Look - Paul Heath

This image was created in the camera by a double exposure. Only one side of the face was illuminated for each exposure.

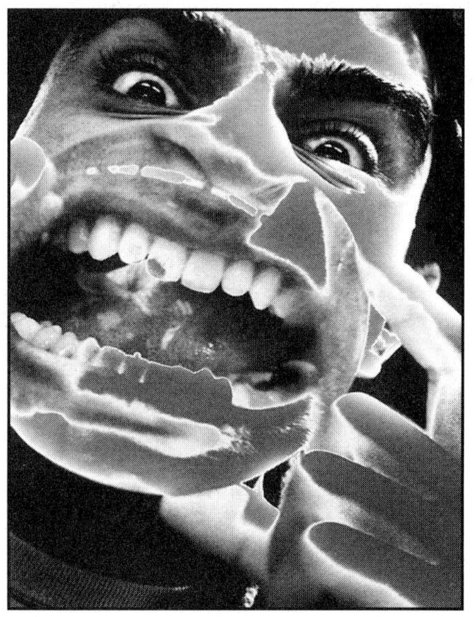

The Mouth - Gareth Neal

This image is a combination of distorting the face using a magnifying lens and then solarising the final print.

Landscape

Mackenzie Charlton

aims

- To increase knowledge of the historical development of the landscape image.
- To express ideas, convictions or emotions through landscape images.
- To develop an understanding of how different techniques can be employed to aid personal expression.

objectives

- **Research** - a broad range of landscape images from both commercial and non-commercial sources.
- **Analyse and evaluate** - the effectiveness of the work you are studying and exchange ideas and opinions with other students.
- **Develop ideas** - produce a study sheet that documents the progress and development of your ideas.
- **Personal response** - produce and present artwork that explores expressive landscape photography in response to the assignment brief in this chapter.

Introduction

Picture postcards, calendars and travel brochures show us glimpses of romantic, majestic and idyllic locations to be admired and appreciated. The beautiful and wonderful are identified, observed, recorded and labelled repeatedly by professionals, tourists and travellers. Mankind is responding to the basic social needs and expectations to capture, document and appreciate. I came, I saw, I photographed. Photography allows the individual to pay homage to beauty and achievement as if in some religious ritual. We mark the occasion of our endeavour and our emotional response by taking a photograph.

> 'Most tourists feel compelled to put the camera between themselves and whatever is remarkable that they encounter. Unsure of other responses they take a picture. This gives shape to experience: stop, take a photograph, and move on.'
>
> Susan Sontag - *On Photography*

In order to avoid a stereotypical representation it is important to connect emotionally with the environment in order to express something personal. How do you as an individual feel about the location and what do you want to say about it?

Mark Galer

Early landscape images

Early landscape images were either created as factual records or looked to the world of painting for guidance in such things as composition and choice of content. Fox Talbot described his early photographs as being created by the 'pencil of nature'. On the one hand the medium was highly valued because of the great respect for nature at this time. On the other hand the medium was rejected as art because many perceived photography as a purely objective and mechanical medium. The question was asked, and is still asked, 'can photography be considered as an artistic medium?'

Although it is the camera that creates the image, it is the photographer who decides what to take and how to represent the subject. This subjective approach enables individuals to express themselves in unique ways whether they use a brush or a camera.

Pictorial photography

The practice of recording the environment as the principal subject matter for an image is a fairly modern concept. Prior to the 'Romantic Era' in the late 18th century, the landscape was merely painted as a setting or backdrop for the principal subject. Eventually the environment and in particular the natural environment began to be idealised and romanticised. The picturesque aesthetic of beauty, unity and social harmony was established by painters such as John Constable and William Turner working just prior to the invention of photography. The first photographic movement was born and was known as '**pictorial photography**'.
Pictorial photographers believed that the camera could do more than simply document or record objectively. The pictorial approach was not so much about information as about effect, mood and technique.
Pictorial photographers often felt, however, that the photographic lens recorded too much detail. This led to photographers employing techniques to soften the final look of the image. These techniques included taking the images slightly out of focus or using print manipulation to remove detail. The aim was to create an image which looked more like a drawing or painting and less like a photograph.

Pictorial style image

Naturalism

Dr Peter Henry Emerson promoted photographic 'Naturalism' in 1889 in his book *Naturalistic Photography for Students of Art*. Emerson believed that photographers shouldn't emulate the themes and techniques of the painters but treat photography as an independent art form. He encouraged photographers to look directly at nature for their guidance rather than painting. He believed that photography should be both true to nature and human vision. Emerson promoted the concept that each photographer could strive to communicate something personal through their work.

Realism

In 1902 a photographer named Alfred Stieglitz exhibited under the title 'Photo-Secessionists' with nonconformist pictorial photographs, choosing everyday subject matter taken with a hand-held camera. These images helped promote photography as an aesthetic medium.

Dunes, Oceano, California 1963 - Ansel Adams

F64

Another member of the group, Paul Strand, pioneered 'straight photography', fully exploring the medium's strengths and careful observation of subject matter. Strand believed that the emphasis should lie in the 'seeing' and not the later manipulation in order to communicate the artist's feelings. The work and ideas influenced photographers such as Edward Weston and Ansel Adams who decided to take up this new 'Realism'. They formed the group F64 and produced images using the smallest possible apertures on large format cameras for maximum sharpness and detail.
Straight photography heralded the final break from the pursuit of painterly qualities by photographers. Sharp imagery was now seen as a major strength rather than a weakness of the medium. Photographers were soon to realise this use of sharp focus did not inhibit the ability of the medium to express emotion and feeling.

Documentary

Photography was invented at a time when the exploration of new lands was being undertaken by western cultures. Photography was seen as an excellent medium by survey teams to categorise, order and document the grandeur of the natural environment. A sense of the vast scale was often established by the inclusion of small human figures looking in awe at the majestic view. These majestic views and their treatment by American photographers contrasted greatly with European landscape photographs. Landscape painters and photographers in Europe did not seek isolation. Indeed seeking out a sense of isolation is problematic in an industrialised and densely populated land.

Bethlehem, Graveyard and Steel Mill - Walker Evans 1935

In the 1930s Roy Stryker of the Farm Security Administration (FSA) commissioned many photographers to document life in America during the depression. Photographers such as Arthur Rothstein, Dorothea Lange and Walker Evans produced images which not only documented the life of the people and their environment but were also subjective in nature.

Activity 1

View the image by Walker Evans on this page and describe what you can actually see (objective analysis) and what you think the image is about (subjective analysis).
Discuss how effective Walker Evans has been in using a landscape image to communicate a point of view.
Can this photograph be considered as Art? Give two reasons to support your answer.

Personal expression

The image can act as more than a simple record of a particular landscape at a particular moment in time. The landscape can be used as a vehicle or as a metaphor for something personal the photographer wishes to communicate. The American photographer Alfred Stieglitz called a series of photographs he produced of cloud formations 'equivalents', each image representing an equivalent emotion, idea or concept. The British landscape photographer John Blakemore is quoted as saying:

> The camera produces an intense delineation of an external reality, but the camera also transforms what it "sees". I seek to make images which function both as fact and as metaphor, reflecting both the external world and my inner response to, and connection with it.

Rocks and Tide, Wales - John Blakemore

> 'Since 1974, with the stream and seascapes, I had been seeking ways of extending the photographic moment. Through multiple exposures the making of a photograph becomes itself a process, a mapping of time produced by the energy of light, an equivalent to the process of the landscape itself.'
>
> John Blakemore 1991

Communication of personal ideas through considered use of design, technique, light and symbolic reference is now a major goal of many landscape photographers working without the constrains of a commercial brief. Much of the art world now recognises the capacity of the photographic medium to hold an emotional charge and convey self-expression.

Alternative realities

There is now a broad spectrum of aesthetics, concepts and ideologies currently being expressed by photographers. The camera is far from a purely objective recording medium. It is capable of recording a photographer's personal vision and can be turned on the familiar and exotic, the real and surreal. This discriminating and questioning eye is frequently turned towards the urban and suburban landscapes the majority of us now live in. It is used to question the traditional portrayal of the rural landscape (romantic and idyllic) as a mythical cliché. It explores the depiction of the natural landscape for many urban dwellers as a mysterious location, viewed primarily through the windscreen of a car and from carefully selected vantage points. Landscape photography can be used to reflect the values of society. The landscape that has been traditionally portrayed as being unified and harmonious may instead be portrayed as confused and cluttered to express the conflict, between expectation and reality.

The Cost of Living - Martin Parr

Photographers also explore their personal relationship with their environment using the camera as a tool of discovery and revelation. To make a photograph is to interact and respond to the external stimuli that surround us. We may respond by creating images that conform to current values and expectations or we may create images that question these values. To question the type of response we make and the type of image we produce defines who we are and what we believe in.

Activity 2

Find two landscape photographs that question social values or act as a metaphor for personal issues that the photographer is trying to express. Discuss whether the communication is clear or ambiguous and how this communication is conveyed.

Expressive techniques

The use of early morning, late afternoon and evening light when the sun is low can be used to increase the range of moods. The changes in contrast and colour can all be used to extend the expressive possibilities.

Recording the sky

Sky may become overexposed when detail is required in the land and land may become underexposed when detail is required in the sky. The problem can be controlled to some extent by filtration, careful exposure in the camera and burning and dodging when printing.

Annapurna

Filters

Polarising filters - these provide increased colour saturation and deeper blue skies when using colour film. The filter is turned when attached to the front of the lens until the desired effect can be seen through the viewfinder. The filter should be removed when the effect is not required. When used in conjunction with a wide-angle lens additional filters (skylight, UV etc.) should be removed. This will eliminate the problem of tunnel vision or clipped corners.

Coloured filters - Yellow, Yellow-Green, Orange and Red filters can all be used with black and white film to help decrease the exposure of a blue sky and obtain detail when printed. Yellow has the least effect and red the most. Underexposure of green colours (fields and trees etc.) is likely when using a red filter so the exposure should be increased to reduce or eliminate the problem.

Composing the landscape

Composing subject matter is more than an aesthetic consideration. It controls the way we read an image and the effectiveness of your communication.

Format and horizon line

The most powerful design elements that you have to work with are choice of format and positioning of the horizon line.

Avoid making the mistake of overusing the vertical or portrait format when faced with tall buildings or trees etc. You should explore both formats when faced with a similar location and compare the communication of each.

The placement of the horizon line within the frame is critical to the final design. A central horizon line dividing the frame into two halves is usually best avoided. Consider whether the sky or foreground is the more interesting element and construct a composition accordingly.

Michelle Greenhalgh

Open or closed landscape

In creating a landscape photograph it is possible to remove the horizon line altogether. By removing the horizon line from the image (through vantage point or camera angle) the photographer creates a '**closed landscape**'. In a closed landscape a sense of depth or scale may be difficult for the viewer to establish. In some instances it is possible that the image may be viewed as an abstract image as the viewer struggles to recognise familiarity through a sense of scale.

Depth

Including foreground subject matter introduces the illusion of depth through perspective and the image starts to work on different planes. Where the photographer is able to exploit lines found in the foreground the viewer's eye can be lead into the picture. Rivers, roads, walls and fences are often used for this purpose.

Solitary Steps

Use of vantage point

By lowering the vantage point or angling the camera down, the foreground seems to meet the camera. A sense of the photographer in, or experiencing, the landscape can be established. In the photograph above the beach and cliff walls are included along with the author's own footsteps to establish a sense of place.

Activity 3

Find examples of photographs created using different formats and compare and discuss the visual effect of each.
Find examples of closed and open landscapes and discuss the different ways we read these images.
Find an example of a photograph where the photographer has created the illusion of depth. How has this been achieved?

The constructed environment

When Arthur Rothstein, FSA photographer, moved a skull a few metres for effect he, and the FSA, were accused of fabricating evidence and being dishonest. A photograph, however, is not reality. It is only one person's interpretation of reality. Rothstein perceived the skull and the broken earth at the same time and so he included them in the same physical space and photograph to express his emotional response to what he was seeing. Is this dishonest? The photograph can act as both a document and as a medium for self-expression. Truth lies in the intention of the photographer to communicate visual facts or emotional feelings. Sometimes it is difficult for a photograph to do both at the same time.

Dreams Will Come True - Matthew Orchard

The majority of photographers are content with responding to and recording the landscape as they find it. A few photographers, however, like to interact with the landscape in a more concrete and active way. Artists such as Andy Goldsworthy use photography to record these ephemeral interactions. Goldsworthy moves into a location without preconceived ideas and uses only the natural elements within the location to construct or rearrange them into a shape or structure that he finds meaningful. The day after the work has been completed the photographs are often all that remain of Goldsworthy's work. The photograph becomes both the record of art and a piece of art in its own right.

Activity 4

Make a construction or arrangement using found objects within a carefully selected public location. Create six images showing this structure or arrangement in context with its surroundings. Consider framing, camera technique and lighting in your approach.

Assignments

Produce six images that express your emotions and feelings towards a given landscape. Each of the six images must be part of a single theme or concept and should be viewed as a whole rather than individually.
People may be included to represent humankind and their interaction with the landscape. The people may become the focal point of the image, but this is not a character study or environmental portrait where the location becomes merely the backdrop.

1. Wilderness.
2. Seascape.
3. Suburbia.
4. Sandscape.
5. Inclement weather.
6. City.
7. Mountain.
8. Industry.
9. Arable land.

Resources

A Collaboration with Nature - Andy Goldsworthy. Abrams. New York. 1990.
On Photography - Susan Sontag. Farrar, Strauss and Giroux. New York. 1989.
Inscape - John Blakemore. Zelda Cheatle Press. London. 1991.
In This Proud Land - Stryker and Wood. New York Graphic Society. New York. 1973.
Land - Faye Godwin. Heinemann. London. 1985.
Naturalistic Photography for Students of the Art - P. H. Emerson. Arno Press. NY. 1973.
The History of Photography: an overview - Alma Davenport. Focal Press. 1991.
The Photograph - Graham Clarke. Oxford University Press. New York. 1997.
The Portfolios of Ansel Adams. New York Graphic Society. Boston. 1977.
The Story of Photography - Michael Langford. Focal Press. Oxford. 1992.

Gallery

Alison Ward

Lucas Dawson

Kalimna Brock

Lucas Dawson

Mark Galer

Portraiture

Mi-Ae Jeong

aims

- ~ To develop an understanding of the genre of photographic portraiture.
- ~ To develop personal skills in directing people.
- ~ To develop essential technical skills to work confidently and fluently with people.

objectives

- ~ **Research** - photographers noted for their skills in photographing people (portraiture and documentary).
- ~ **Analyse and evaluate** - the effectiveness of the work you are studying and exchange ideas and opinions with other students.
- ~ **Discussion** - exchange ideas and opinions with other students.
- ~ **Personal response** - produce images through close observation and selection that demonstrate both a comfortable working relationship with people at close range and appropriate design and technique.

Introduction

The craft of representing a person in a single still image or 'portrait' is to be considered a skilled and complex task. The photographic portrait (just as the painted portrait that influenced the genre) is not a candid or captured moment of the active person but a crafted image to reveal character.
The person being photographed for a portrait should be made aware of the camera's presence even if they are not necessarily looking at the camera when the photograph is made. This requires that the photographer connect and communicate with any individual if the resulting images are to be considered portraits. Portraits therefore should be seen as a collaborative effort on the part of the photographer and subject. A good photographic portrait is one where the subject no longer appears a stranger.

Matthew Orchard

The physical surroundings included in a portrait offer enormous potential to extend or enhance the communication. Just as facial expression, body posture and dress are important factors, the environment plays a major role in revealing the identity of the individual.

Activity 1

Look through assorted books, magazines and newspapers and collect four portrait photographs. The environment should be a key feature in two of the four images.
Describe the subject's character in each of the images.
What can you see within each image that leads you to these conclusions about the subjects' character.

Design

Choosing a suitable background or backdrop for the portrait can greatly influence the final design of the image. Using a plain backdrop with limited detail can retain focus on the individual being photographed whilst choosing an informative location can extend the communication and design possibilities. If the photographer is to reveal any connection between the subject and the background the two elements must be carefully framed together. Vantage point and the relationship and connection between foreground and background become major design considerations for the portrait photograph.

Shaun Guest

Format

The choice of vertical or horizontal framing and the placement of the subject within the frame will effect the quantity of the background that can be viewed in the image. A centrally placed subject close to the camera will limit the background information. This framing technique tends to be overused but should not be ruled out for creating successful portraits. The horizontal format is common when creating environmental portraits. Using the portrait format for environmental portraits usually requires the photographer to move further back from the subject so that background information is revealed.

Activity 2

Find four portraits that demonstrate the different ways a photographer has framed the image to alter the design and content.

Discuss the vantage point, depth of field and subject placement in all of the images.

Composing two or more people

Composing two or more people within the frame for a portrait can be difficult. The physical space between people can become very significant in the way we read the image. For a close-up portrait of two people the space between them can become an uncomfortable design element. Careful choice of vantage point or placement of the subjects is often required to achieve a tight composition making optimum use of the space within the frame.

Kata Bayer

The situation most often encountered is where two people sit or stand side by side, shoulder to shoulder. If approached face on (from the front) the space between the two people can seem great. This can be overcome by shooting off to one side or staggering the individuals from the camera. The considerations for design are changed with additional subjects.

Activity 3

Collect four portrait images with two to five subjects.
In at least one image the subject should have been placed in the foreground.
Comment on the arrangement of the subjects in relation to the camera and the effectiveness of the design.

Depth of field

Sophisticated 35mm SLR cameras often provide a 'Portrait Mode'. When this programme mode is selected a combination of shutter speed and aperture is selected to give the correct exposure and a visual effect deemed suitable for portrait photography by the camera manufacturers. The visual effect aimed for is one where the background is rendered out of focus, i.e. shallow depth of field. This effect allows the subject to stand out from the background, reducing background information to a blur. Although this effect is appropriate for many portrait images it is not suitable for most environmental portraits where more information is required about the physical surroundings and environment. The student intending to create portraits is recommended to use their camera in either fully manual or aperture priority mode so that maximum control is maintained.

Buddhist Temple, Singapore

Appropriate focal length

Photographic lenses can be purchased by manufacturers which are often referred to as 'portrait lenses'. The 'ideal' portrait lens is considered by the manufacturers to be a medium telephoto lens such as a 135mm lens for a 35mm camera. This lens provides a visual perspective that does not distort the human face when recording head and shoulder portraits. The problem of distortion, however, is not encountered with shorter focal length lenses if the photographer is not working quite so close to the subject. To record environmental portraits with a telephoto lens would require the photographer to move further away from the subject and possibly lose the connection with the subject that is required. Standard and wide-angle lenses are suitable for environmental portraiture.

Activity 4

Photograph the same subject varying both the depth of field and focal length of the lens. Discuss the visual effects of each image.

Revealing character

Significant and informative details can be photographed with the subject. These details may naturally occur or be introduced for the specific purpose of strengthening the communication. Connections may be made through the 'tools of the trade' associated with the individual's occupation. Informative artefacts such as works of art or literature may be chosen to reflect the individual's character. Environments and lighting may be chosen to reflect the mood or state of mind of the subject.

Ann Ouchterlony

The objects or subject matter chosen may have symbolic rather than direct connection to the subject. In the above image the bent walking stick of the old man and the path travelled could be seen to represent the journey of life.

Activity 5

Find one portrait image that has included significant or informative detail.
Describe the importance of the additional information and how it is likely to be read by the viewer.

Photographing strangers

It can sometimes feel awkward or embarrassing to be either photographed or photograph someone at close range. These feelings of awkwardness, embarrassment or even hostility may arise out of the subject's confusion or misinterpretation over the intent or motive of the photographer.
The photographer's awkwardness or reticence to introduce themselves to the subject often comes from the fear of rejection. Never be tempted to think that a portrait can be created using a telephoto lens with the subject unaware. Although some interesting images of people are captured this way they are not strictly portraits.

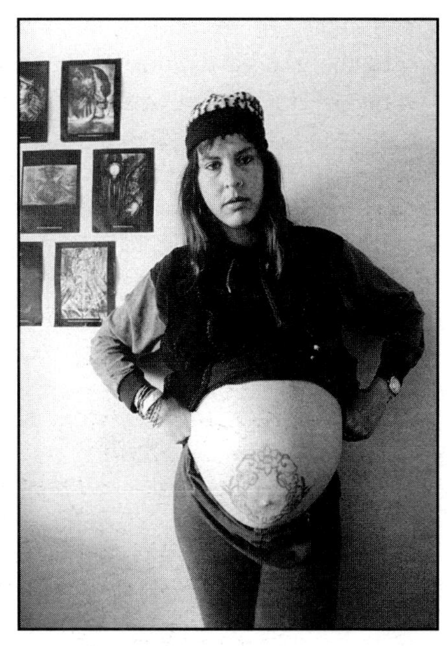

Stephen Rooke

Interaction

Putting a subject at ease in front of the camera is dependent on two main factors.

1. **The subject is clear about the photographers *motive*.**
2. **The subject sees *value* in the photographs being made.**

Motive

Many people view a photographer with curiosity or suspicion. Who is the photographer and why are they taking photographs? It is essential that the photographer learns to have empathy with the people he/she intends to photograph. A brief explanation is therefore necessary to help people understand that your intentions are harmless.

Value

Many people see the activity or job that they are doing as uninteresting or mundane. They may view their physical appearance as non-photogenic. You should explain to the subject what you find interesting or of value and why. If the activity the subject has been engaged in appears difficult or demanding and requires skill, patience or physical effort, you should tell them. You should continue to ask questions whilst photographing so that the subject is reassured that your interest is genuine.

Directing the subject

The photographer should display an air of confidence and friendliness whilst directing subjects. Subjects will feel more comfortable if the photographer clearly indicates what is expected of them. There can be a tendency for inexperienced photographers to rush a portrait. The photographer may feel embarrassed, or feel that the subject is being inconvenienced by being asked to pose. The photographer should clarify that the subject does have time for the photograph to be made and indicate that it may involve more than one image being created. A subject may hear the camera shutter and presume that one image is all that is required.

Passive subject

Ask your subject to pause from any activity that they are engaged in. You can remain receptive to the potential photographic opportunities by keeping the conversation focused on the subject and not on oneself.

Simon Sandlant

Expression and posture

Often a subject will need reminding that a smile may not be necessary. Subjects may need guidance on how to sit or stand, what they should do with their hands and where to look. It may be a simple case of just reminding them how they were standing or sitting prior to having their photograph taken. Have a few poses memorised so that you can position somebody who is feeling or looking awkward.

Shooting decisively

As you take longer to take the picture the subject will often feel more and more uncomfortable about their expression and posture. To freeze human expression is essentially an unnatural act. If the camera is to be raised to the eye to capture the image (as opposed to using a tripod), exposure, framing and focus should all be considered first.

Activity 6

Direct two individuals towards a relaxed expression and body posture.
Discuss the process of direction for each.

Character study

A series of portraits may be taken around a single character, or characters, connected by profession, common interest or theme.
With additional images it is possible to vary the content and the style in which the subject is photographed to define their character within the study. The photographer may choose to include detail shots such as hands or clothing to increase the quality of information to the viewer. The study may also include images which focus more on the individual (such as a straight head and shoulders portrait) or the environment to establish a sense of place.

The Big Issue - Sean Killen

The images above show the diversity of approach to present the character of a single individual. The images are of Martin, a homeless individual who lived under a bridge in Melbourne and who sold copies of *The Big Issue* to support himself.

Activity 7

Collect one character study where the photographer has varied the content of the images to define the character or characters of the individual or individuals.
Describe the effectiveness of the additional images that are not portraits.

Assignment

Produce six portraits giving careful consideration to design, technique and communication of character. At least one of the images should include more than one person. Choose one category from the list below.

1. Manual labourers ....dockers, builders, mechanics, bakers etc.
2. Professional people .... doctors, nurses, lawyers etc.
3. Craftsmanship ....potters, woodworkers, violin makers etc.
4. Club or team members .... cricketers, golfers, scouts etc.
5. Family and friends ....different generations etc.
6. Character study....................celebrity, politician, busker etc.
   This study should include portraits, environmental portraits and close-up detail.

Resources

Editorial:
National Geographic
Newspapers and magazines

Photographers:
Bill Brandt, Brian Griffin, Sebastio Salgado.
East 100th Street - Bruce Davidson. Harvard University Press. Cambridge. 1970.
Pictures of People - Nicholas Nixon. Museum of Modern Art. New York. 1988.
Men Without Masks - August Sander. New York Graphic Society. New York. 1973.
One Mind's Eye - Arnold Newman. Secker and Warburg. London. 1974.

Gallery

Chris Augustnyk

Bec McCubbin

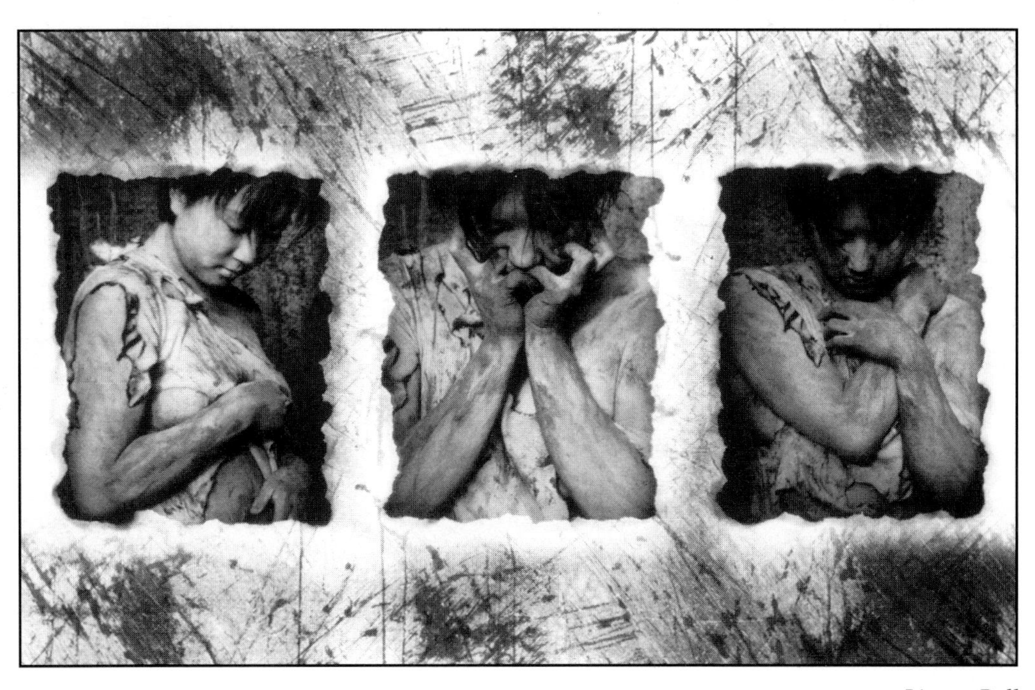

Lizette Bell

Lizette Bell

Shaun Guest

Photo-story

Andrew Goldie

aims

- To understand visual communication through narrative techniques.
- To develop an awareness of receptive and projective styles of photography.
- To understand the process of editing to clarify or manipulate communication.
- To increase awareness of commercial, ethical and legal considerations.

objectives

- **Research** - photo-stories and discuss the communication, narrative and structure.
- **Analyse and evaluate** - the effectiveness of the work you are studying and exchange ideas and opinions with other students.
- **Develop ideas** - produce a study sheet that documents the progress and development of your ideas.
- **Personal response** - capture and/or create images to communicate a specific narrative and edit the work collaboratively.

Introduction

The purpose of constructing a photographic essay is to communicate a story through a sequence of images to a viewer. Just as in writing a book, a short story or a poem the photographer must first have an idea of what they want to say. In a photographic essay it is the images instead of words that must be organised to tell the story. Individual images are like descriptive and informative sentences. When the images are carefully assembled they create a greater understanding of the individual, event or activity being recorded than a single image could hope to achieve. Words should be seen as secondary to the image and are often only used to clarify the content.

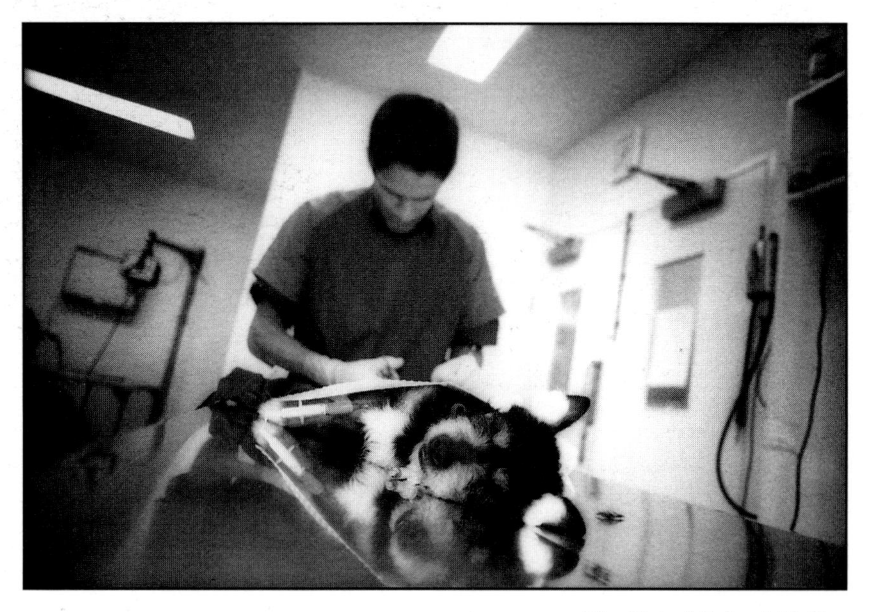

The Vet - Michael Mullan

The first stories

In 1890 the photographer Jacob Riis working in New York produced one of the earliest photographic essays titled *How the other half lives*. *National Geographic* magazine began using photographs in 1903 and by 1905 they had published an eleven-page photographically illustrated piece on the city of Lhassa in Tibet. In 1908 the freelance photographer Lewis W. Hine produced a body of work for a publication called *Charities and the Commons*. The photographs documented immigrants in the New York slums. Due to the concerned efforts of many photographers working at this time to document the 'human condition' and the public's growing appetite for the medium, photography gradually became accepted. The first 'tabloid newspaper' (the *Illustrated Daily News*) appeared in the USA in 1919. By this time press cameras were commonly hand held and flash powder made it possible to take images in all lighting conditions.

FSA

Migrant Mother - Dorothea Lange

The 1930s saw a rapid growth in the development of the photographic story. The small 35mm Leica camera began to be mass produced. It was equipped with a fast lens and would take advantage of the high quality 35mm roll film developed for the movie industry.

During this decade the Farm Security Administration (FSA) commissioned photographers to document America in the grip of a major depression. Photographers including Russell Lee, Dorothea Lange, Ben Shahn, Carl Mydans, Walker Evans and Arthur Rothstein took many thousands of images over many years. This project provides an invaluable historical record of culture and society whilst developing the craft of documentary photography.

The photo agencies

In the same decade that the FSA was operating *Life* magazine was born and this saw a proliferation of like-minded magazines such as *Picture Post* in the UK. These publications dedicated themselves to showing 'life as it is'. Photographic agencies were formed in the 1930s and 1940s to help feed the public's voracious appetite for news and entertainment. The greatest of these agencies Magnum was formed in 1947 by Henri Cartier-Bresson, Robert Capa, Chim (David Seymour) and George Rodger. Magnum grew rapidly with talented young photographers being recruited to their ranks. The standards for honesty, sympathetic understanding and in-depth coverage were set by such photographers as W. Eugene Smith. Smith was a *Life* photographer who produced extended essays staying with the story until he felt it was an honest portrayal of the people he photographed. In 1954 *Life* printed 'A man of mercy', a twenty-five image story that Smith had created about Albert Schweitzer. Smith felt that his story had been edited too heavily and resigned. He went on to produce the book *Minamata* about a small community in Japan who were being poisoned by toxic waste being dumped into the waterways where the people fished. This was and remains today an inspirational photo-essay.

Activity 1

Research a photographic story that was captured by either an FSA photographer or a photographer working for the Magnum photo agency.

What do the images communicate about the human condition?

Visual communication

Photographic stories are the visual communication of personal experience, as such each story is potentially unique and is the ideal vehicle for personal expression. To communicate coherently and honestly the photographer must connect with what is happening. To connect the photographer should research, observe carefully, ask questions and clarify the photographer's personal understanding of what is happening. Unless the photographer intends to make the communication ambiguous it is important to establish a point of view or have an 'angle' for the story. This can be achieved by acknowledging feelings or emotions experienced whilst observing and recording the subject matter. All images communicate and most photographers aim to retain control of this communication. Photography can be used as a powerful tool for persuasion and propaganda and the communication of content should always be the primary consideration of the photographer.

Kim Noakes

Choosing a subject

The most popular subject for the photographic story has always been the 'human condition'. This is communicated through experience-based discovery. The aim is to select one individual or group of individuals and relate their story or life experience to the viewer. The story may relate the experience of a brief or extended period of time.

Finding a story, gaining permission to take images and connecting with the individuals once permission has been granted are some of the essential skills required to produce a successful story. Tracking down a story often requires curiosity, perseverance, motivation and patience. These skills are required by the majority of professional photojournalists who are freelance. Freelance photographers find, document and sell their own stories.

The comfort zone

The 'comfort zone' is a term used to describe the familiar surroundings, experiences and people that each of us feel comfortable in and with. They are both familiar and undemanding of us as individuals. Photography is an ideal tool of exploration which allows us to explore environments, experiences and cultures other than our own. For professional photojournalists this could be attending the scene of a famine or a Tupperware party.

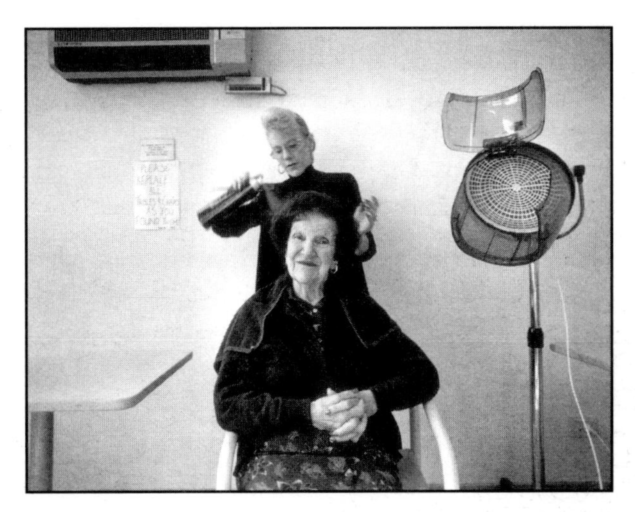

Hairspray - Michael Davies

The student photographer may feel they have to travel great distances in order to find an exotic or unusual story. Stories are, however, much closer at hand than most people realise. Interesting stories surround us. Dig beneath the surface of any seemingly bland suburban population and the stories will surface. People's triumphs, tragedies and traumas are evolving every day, in every walk of life. The interrelationships between people and their environment and their journey between birth and death are the never ending, constantly evolving resource for the documentary photographer. The photographer's challenge is to find and connect in a non-threatening and sympathetic way to record this. The student of photojournalism should strive to leave their personal 'comfort zone' in order to explore, understand and document the other. The photographer should aim to become proactive rather than waiting to become inspired and find out what is happening around them.

Activity 2

Find two meaningful photographic stories containing at least four images.
What is being communicated in each story?
Have the captions influenced your opinion about what is happening?
Could a different selection of images alter the possible communication?
Make a list of five photo-essays you could make in your own home town.
Describe briefly what you would hope to find out and communicate with your images.

Capturing a story

How many movie films have you seen where the opening scene begins with a long and high shot of a town or city and moves steadily closer to isolate a single street or building and then a single individual. This gives the viewer a sense of the place or location that the character inhabits. A story constructed from still images often exploits the same technique. To extend and increase the communication of a series of images the photographer should seek to vary the way in which each image communicates. There is a limit to the communication a photographer can achieve by remaining static, recording people from only one vantage point. It is essential that the photographer moves amongst the people exploring a variety of distances from the subject. Only in this way will the photographer and the viewer of the story fully appreciate and understand what is happening.

The photographer should aim to be a witness or participant at an activity or event rather than a spectator.

Vietnam Vets at the Shrine of Remembrance, Melbourne

The images that create a well-crafted photographic story can usually be divided or grouped into four main categories. Not all stories contain images from all four categories but many editors expect to see them. The categories are:

1. Establishing image.
2. Action image.
3. Portrait.
4. Detail image.

Establishing image

In order to place an event, activity or people in context with their environment it is important to step back and get an overview. If the photographer's essay is about a small coal-mining community in a valley, the viewer needs to see the valley to get a feeling for the location. This image is often referred to as the establishing image but this does not necessarily mean that it is taken or appears first in the story. Often the establishing image is recorded from a high vantage point and this technique sets the stage for the subsequent shots. In many stories it can be very challenging to create an interesting establishing image. An establishing image for a story about an animal refuge needs to be more than just a sign in front of the building declaring this fact. The photographer may instead seek out an urban wasteland with stray dogs and the dog catcher to set the scene or create a particular mood.

Coal mine in the Rhondda valley

Action image

This category refers to a medium-distance image capturing the action and interaction of the people or animals involved in the story. Many of the images the photographer captures may fall into this category, especially if there is a lot happening. It is, however, very easy to get carried away and shoot much more than is actually required for a story to be effective. Unless the activity is unfolding quickly and a sequence is required the photographer should look to change the vantage point frequently. Too many images of the same activity from the same vantage point are visually repetitive and will usually be removed by an editor.

Portraits

Portraits are essential to any story because people are interested in people and the viewer will want to identify with the key characters of the story. Unless the activity the characters are engaging in is visually unusual, bizarre, dramatic or exciting the viewer is going to be drawn primarily to the portraits. The portraits and environmental portraits will often be the deciding factor as to the degree of success the story achieves. The viewer will expect the photographer to have connected with the characters in the story and the photographs must illustrate this connection. The connection is aided by the use of standard and wide-angle lenses which encourage an interactive working distance.

Portraits may be made utilising a variety of different camera distances. This will ensure visual interest is maintained. Environmental portraits differ from straight head and shoulder portraits in that the character is seen in the location of the story. The interaction between the character and their environment may extend the communication beyond making images of the character and location separately.

Roly Imhoff

Close-up or detail image

The final category requires the photographer to identify significant detail within the overall scene. The detail is enlarged either to draw the viewer's attention to it or to increase the amount or quality of this information. The detail image may be required to enable the viewer to read an inscription or clarify small detail. The detail image in a story about a craftsperson who works with their hands may be the hands at work, the fine detail of the artefact they have made or an image of a vital tool required for the process. For very small detail the photographer may require a macro lens or magnifying dioptres that can be attached to the existing lens. When the detail image is included with images from the other three categories visual repetition is avoided and the content is clearly communicated.

Activity 3

Find one photographic story that contains images from all four categories listed.
How does each image contribute to the story?
Describe alternative images for each of the categories in the story you have chosen?

Creating a story

To create a story it is possible to plan and arrange the timing, the subject matter and location so that the resulting images fit your requirements for the communication of a chosen narrative.

To create a story you first need to start with the concept or idea that you would like to illustrate. The story can be inspired by the words of a poem or novel that you have read or the lyrics of a song. Start with words that create images for the mind and then illustrate what is in your minds' eye. Keep the story short or simple and remember to be realistic about your technical ability and the resources (human and physical) to create the story for the camera.

Sharounas Vaitkus

A '**storyboard**' should first be made that shows the sequence of events that can later be photographed. The categories used in '**Capturing a story**' can be used effectively to create a story as well. This will ensure that all of the images required to communicate the story are made with efficient use of your resources.

Activity 4

Look at the work of a photographer who uses sequences of images to narrate a story.
Can the story be read without words?
Discuss the effectiveness of the communication.

Editing a story that has been captured

The aim of editing work is to select a series of images from the total production to narrate an effective story. Editing can be the most demanding aspect of the process requiring focus and energy. The process is a compromise between what the photographer originally wanted to communicate and what can actually be said with the images available.

Editorial objective

The task of editing is often not the sole responsibility of the photographer. The task is usually conducted by an editor or in collaboration with an editor. The editor considers the requirements of the viewer or potential audience whilst selecting images for publication. The editor often has the advantage over the photographer in that they are less emotionally connected to the content of the images captured. The editor's detachment allows them to focus on the ability of the images to narrate the story, the effectiveness of the communication and the suitability of the content to the intended audience.

The process

Images are viewed initially as a collective. A typical process is listed as follows:

- All the proof sheets, prints or transparencies are spread out on a table or light box.
- All of the visually interesting images and informative linking images are selected and the rest are put to one side.
- Similar images are grouped together and the categories (establishing, action, portrait and detail) are formed.
- Images and groups of images are placed in sequence in a variety of ways to explore possible narratives.
- The strength of each sequence is discussed and the communication is established.
- Images are selected from each category that reinforce the chosen communication.
- Images that contradict the chosen communication are removed.
- Cropping and linking images are discussed and the final sequence established.

Activity 5

Take two rolls of film of a chosen activity or event taking care to include images from all of the four categories discussed in this chapter.
Edit the work with another student who can take on the role of an editor.
Edit and sequence the images to the preference of the photographer and then again to the preference of the editor.
Discuss the editing process and the differences in outcome of both the photographer's and editor's final edit.

Ethics and law

Will the photograph of a car crash victim promote greater awareness of road safety, satisfy morbid curiosity or just exploit the family of the victim? If you do not feel comfortable photographing something, question why you are doing it. A simple ethical code of practice used by many photographers is: **'The greatest good for the greatest number of people'**. Papparazzi photographers hassle celebrities to satisfy public curiosity and for personal financial gain. Is the photographer or the public to blame for the invasion of privacy?
The first legal case for invasion of privacy was filed against a photographer in 1858. The law usually states that a photographer has the right to take a picture of any person whilst in a public place so long as the photograph:

- is not used to advertise a product or service;
- does not portray the person in a damaging light (called deformation of character).

If the photographer and subject are on private property the photographer must seek the permission of the owner. If the photograph is to be used for advertising purposes a model release should be signed.
Other legal implications usually involve the sensitivity of the information recorded (military, political, sexually explicit etc.) and the legal ownership of copyright. The legal ownership of photographic material may lie with the person who commissioned the photographs, published the photographs or the photographer who created them. Legal ownership may be influenced by country or state law and legally binding contracts signed by the various parties when the photographs were created, sold or published.

Digital manipulation

Images are often distributed digitally which allows the photographer or agency to alter the images subtly in order to increase their commercial potential. Original negatives or transparencies may never be sighted by an editor. A photographer, agency or often the publication itself may enhance the sharpness, increase the contrast, remove distracting backgrounds, remove information or combine several images to create a new image. What is legally and ethically acceptable is still being established in the courts of law. The limits for manipulation often rests with the personal ethics of the people involved.

Activity 6

Discuss the ethical considerations of the following:

- Photographing a house fire where there is the potential for loss of life.
- Photographing a celebrity, who is on private property, from public property.
- Manipulating a news image to increase its commercial viability.

Assignments

Produce a six image photo-story giving careful consideration to communication and narrative technique. For each of the five assignments it is strongly recommended that students:

- ~ Choose activities, events or social groups that are repeatedly accessible.
- ~ Avoid choosing events that happen only once or run for a short period of time.
- ~ Approach owners of private property in advance to gain relevant permission.
- ~ Have a back-up plan should permission be denied.
- ~ Introduce yourself to organisers, key members or central characters of the story.
- ~ Prepare a storyboard if you are creating a story.

Capture a story. Document one of the following categories:

1. Manual labour or a profession.
2. Minority, ethnic or fringe group in society.
3. An aspect of modern culture.
4. An aspect of care in the community.

Create a story. Illustrate one of the following:

5. The dream
6. The journey
7. The encounter
8. A poem, short story or the lyrics to a song.

Resources

Photofile - Duane Michals. Thames and Hudson. London. 1990.
Farewell to Bosnia - Gilles Peress. Scalo. New York. 1994.
In This Proud Land - Stryker and Wood. New York Graphic Society. New York. 1973.
Minamata - W. Eugene and Aileen Smith. Chatto and Windus. London. 1975.
Sleeping with Ghosts - Don McCullin. Vintage. London. 1995.
The Concerned Photographer - Cornell Capa. Thames and Hudson. London. 1972.
Workers - Sebastio Salgado. Phaidon Press. London. 1993.

National Geographic magazine.
American Photo magazine.

Black Star photo agency - http://www.blackstar.com
FSA web site - http://www.memory.loc.gov/ammem/fsowhome.html

Gallery

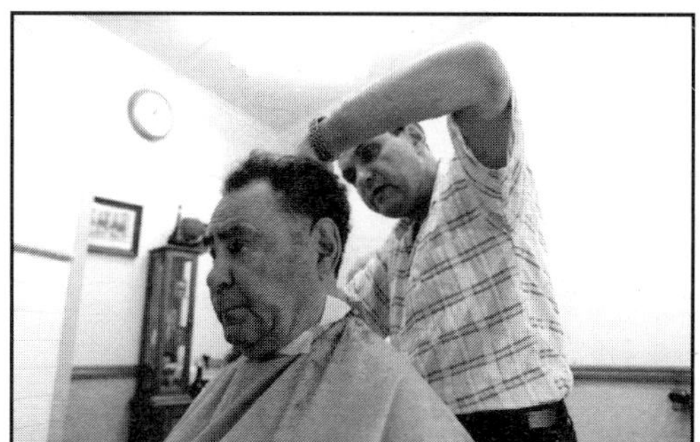

Michael Davies

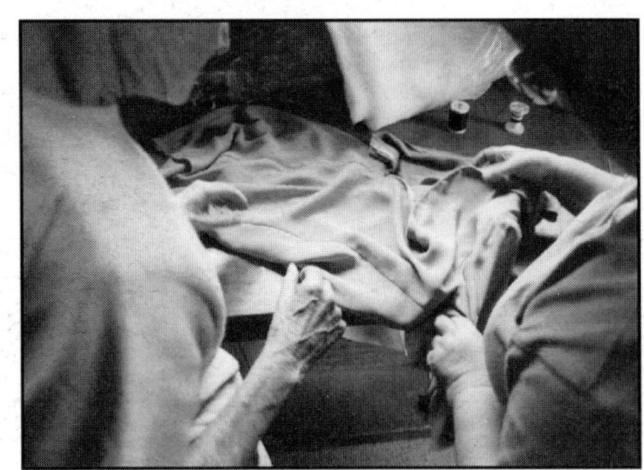

Anthony Secatore

Visual Literacy

Mark Galer

aims

- To develop an understanding of why and how photographic images are constructed.
- To develop an awareness of how the photographic image can be manipulated to communicate specific messages.

objectives

- **Research** - a range of photographic images from the printed media, documentary and advertisements.
- **Analyse and evaluate** - how effective the media is in its aims. Exchange ideas and opinions with other students.
- **Develop ideas** - produce a study sheet that documents the progress and development of your ideas.
- **Personal response** - produce artwork individually and in teams to demonstrate how photography can be used to communicate different messages.

Introduction

A photograph, whether it appears in an advertisement, a newspaper or in a family album is often regarded as an accurate and truthful record of real life. Sayings such as 'seeing is believing' and 'the camera never lies' reinforce these beliefs. In this study guide you will learn that the information we see in photographs is often carefully selected by the photographer or by the editor so that what we believe we're seeing is usually what somebody else would like us to see. By changing or selecting the information we can change the message.
The concepts of time, motion and form that exist in the real world are accurately translated by the photographic medium into timeless and motionless two-dimensional prints. Photographers, editors and the general public frequently use photography to manipulate or interpret reality in the following ways:

1. Advertisers set up completely imaginary situations to fabricate a dream world to which they would like us to aspire.
2. News editors choose some aspects of an event, excluding others, to put across a point of view.
3. Family members often choose to record only certain events for display in the family album, excluding others, so that we portray the image of the family as a happy and united one.

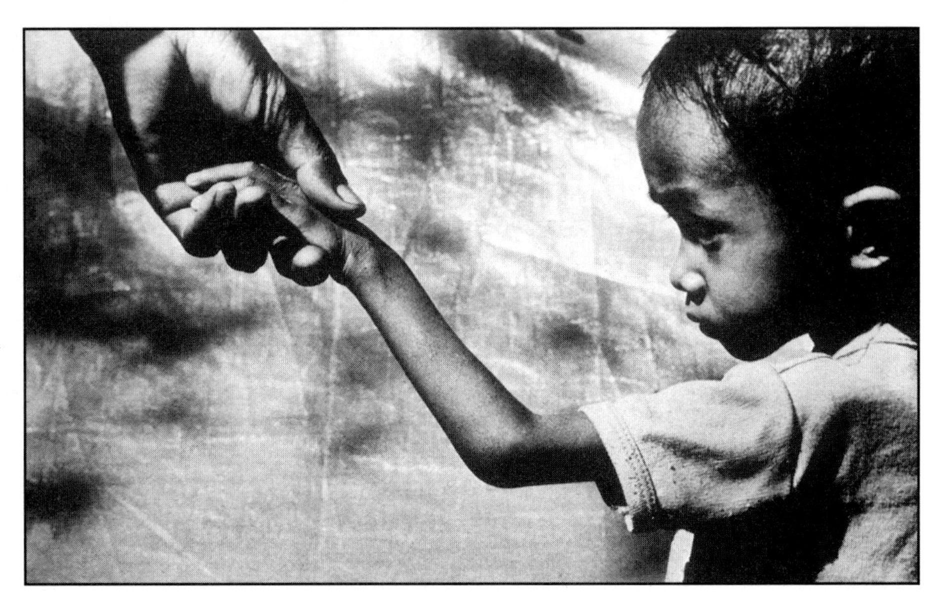

Cambodian Refugee Camp - Burk Uzzle

Photography is a powerful media tool capable of persuasion and propaganda. It appears to offer truth when in reality it can portray any manipulative or suggestive statement.
The camera may record accurately but it is people who choose what and how it records. A photograph need only be sufficiently plausible so that it appears to offer the truth.

Manipulation techniques

Framing and cropping - This technique is used to include or exclude details that may change the meaning of the photograph. The photographer, by moving the frame, or the editor by cropping it, defines the content of what we see. By framing two facts it can create a relationship where none may otherwise exist. By excluding a fact it may break one, e.g. an old woman sitting on a park bench feeding the birds might appear lonely if we have moved the frame slightly to exclude her granddaughter playing nearby.

Editing - A photographer chooses a subject matter that he believes to be important and selects the decisive moment to take the picture. In order to present a personal point of view the editor may then decide to select only one aspect of the photographer's work.

Captions - A caption may give the picture a moral, social, political, emotional, or historical meaning that emphasises the intended message.

Focusing - The amount we see 'in-focus' can vary and draw our attention to a specific part of the photograph.

Vantage point or angle - The photograph may be taken from above, below or at eye level. Photographs taken from a low vantage point tend to increase the power and authority of the subject whilst those taken from above tend to reduce it.

Lighting - This can change the atmosphere or mood of the scene, e.g. soft and subtle, dramatic or eerie.

Film type - Black and white or colour. The contrast and colours may vary.

Subject distance - The closer we appear to be to the subject the more involved we become and the less we discover about the subject's environment or location.

Lens distortion - By using a telephoto lens detail and depth can be compressed, condensed and flattened. Subject matter appears to be closer together. By using a wide-angle lens distances are exaggerated and scale is distorted. Things that are close to the lens look larger in proportion to their surroundings than they really are. Things in the distance look much further away. An estate agent may use a wide angle lens to make a room look bigger.

Composition - This is a technique which is used to describe how lines, shapes and areas of tone or colour are placed within the picture frame to attract or guide our attention, e.g. diagonal lines, whether real or implied, make the picture more dramatic and give us a sense of movement. Due to the strong colour contrast, a person wearing a red jacket in a green field will instantly draw our attention.

Photographic categories

Documentary - A factual record.
Advertising - The promotion of a product.
Art - A display of skill and/or creative expression.

Activity 1 - coverage

Collect one glossy magazine, one 'tabloid' newspaper and one 'broadsheet' newspaper.
What percentage of each publication is covered by photographs?
What percentage of these are advertising images?

Activity 2 - analysis

1. Remove a selection of photographs complete with captions from the media sources you have used in activity 1. After studying each photograph prepare a table, as in the example below, using the same headings. In describing the subject matter you will need to consider:

~ Who is the main figure in the photograph.
~ How they look.
~ Who else appears and how they react to the main figure.
~ What else appears in the photograph e.g. objects, location, setting.

Category	Subject matter	Photographic techniques	Implied message
Advert for beauty product.	Woman well-dressed in the grounds of a large country house. Handsome, well-dressed young man looking on.	Warm colours, softly lit low vantage point, focus on woman. Woman placed centrally in foreground.	Affluence, glamour, sexual admiration, happiness, contentment.

2. Twenty years ago John Berger wrote in *Ways of Seeing* that advertising showed us images of ourselves, made glamorous by the products it was trying to sell. He claims the images make us envious of ourselves as we might become, as a result of purchasing the product. Berger also believes that advertising makes us dissatisfied with our present position in society.
Look at the images you have collected. Which of the following are we likely to envy or admire in each image:

~ Power
~ Prestige and status
~ Happiness
~ Glamour
~ Sexuality

Do any of your advertisements appeal to different emotions other than admiration and envy? With advertising becoming increasingly sophisticated and less blatant, do John Berger's claims still ring true?

Activity 3 - captions

In this activity you will look at how captions reinforce the message implied in a photograph. Captions that accompany photographs can be:

- ~ Explicit - stating something very clearly as fact.
- ~ Implicit - suggesting that something is true.

For this activity you need to choose photographs and their captions from both the advertising and documentary categories. Rewrite the captions that accompany each to give a different point of view or bias. You may like to think of captions that change the moral, political, emotional or historical meaning of the photograph. The caption you choose may even change the category in which the photograph first appeared.
Present your work and discuss how you have changed the meaning of the photographs and how effective the new messages have become.

Activity 4 - juxtaposition

Photomontage is a technique where separate photographs are combined to create new meanings. The interaction between the new elements creates a new meaning. Below we see the unintentional and incongruous juxtaposition of two posters in the street.

> An entirely new meaning is made in the distance that exists between the glamour and eroticism of western fashion and the economics of survival in the third world.
>
> Peter Kennard - *Photomontage Today.*

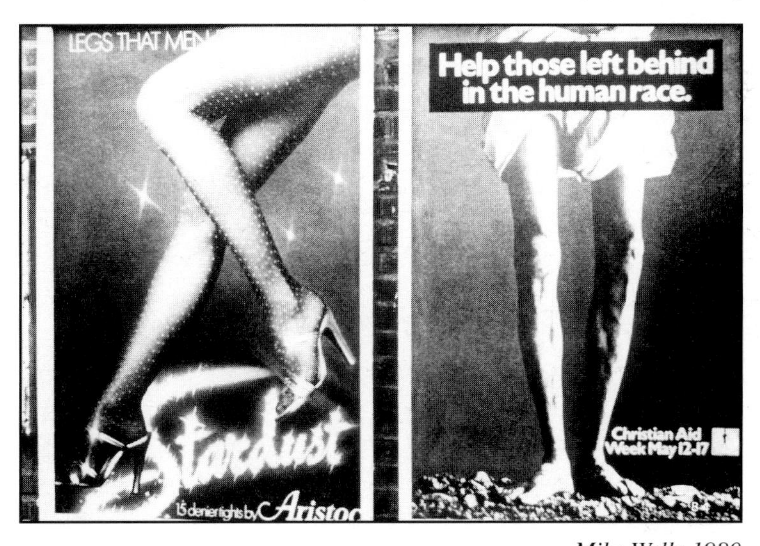

Mike Wells 1980

Using some of the photographs you have collected, cut out various elements from each and rearrange them to make new images with new messages. The sort of new images that work best are where strong contrasts are placed together, e.g. unemployment queues standing next to advertisements for luxury merchandise.

Activity 5 - group practical

For this practical activity you should split up into groups of three or four. Your group will have a broader perspective if the members are not all male or female.
The things you will need for this activity are as follows:

- 35mm SLR camera
- 24 exposure roll of film
- location
- assortment of props

Using appropriate photographic techniques as discussed earlier, take photographs for each of the following categories:

- Advertising
- Documentary

Repeat the exercise using one or more inappropriate techniques to create a satirical photograph for each of the categories. Allow six frames for each individual in the group.

Important

Discuss in detail each shot before you commit yourself to taking the picture. You might like to consider a possible caption for each shot before planning your photographs.
Prepare for any technical difficulties that you think you may encounter, e.g. working with a tripod if you are shooting indoors without a flash.
You have probably already noticed how many photographers fill the frame with their subject matter in order to make a bold statement. Avoid standing too far back in your own shots unless you have a specific reason to do so.

Activity 6 - editing

This activity requires that you work in pairs to act as a newspaper editor and photojournalist. Your objective is to produce a short news story that is either **biased** or **unbiased**.
Use titles, photographs and captions. Text is optional but you may consider pasting down some text to create a realistic mock-up of the finished article.
The editor may choose not to tell the photographer which stance he or she is going to take on the subject. The photographer may choose to manipulate the range of images supplied to the editor.
Your teacher will give you advice on the range of topics you may choose from, or you can make your own suggestions.

Activity 7 - the family album

1. In this activity you will be looking at the way the family is portrayed both in the media and in our own family albums.
Collect some media photographs that portray the family and analyse them using the criteria you used in activity 2:

- ~ Category
- ~ Subject matter
- ~ Photographic techniques
- ~ Implied message

Apply the same criteria to the photographs that appear in a typical family photograph album. Pay particular attention to the way the photographs have been edited. What aspects of family life were edited out or simply not recorded in the first place? Do you think the family album is a true representation of family life? How useful or damaging would it be to alter the type of events that are considered worthy of being placed in the family album?

2. Write a short essay evaluating your findings. Discuss how you think media images have affected you, your family and the general public. Try to be as honest as you can.

Resources

The following resources are suggestions only. You or your teacher may to add to this list.

Images from the press

Newspapers (both 'tabloid' and 'broadsheet').
Magazines.

Recommended reading

On Photography - Susan Sontag. Farrar, Stauss and Giroux. New York. 1989.
Ways of Seeing - J. Berger. Penguin Books. London. 1972.

Images with accompanying text

About 70 Photographs - C. Steele-Perkins. Arts Council of Great Britain.
Basic Photography - M. Langford. Focal Press. Oxford. 1997.
Looking at Photographs - J. Szarkowski. Museum of Modern Art, New York.
Media Studies - B. Dutton. Longman. 1989.
Images for the End of the Century - P. Kennard. Journeyman Press. London. 1990.

Video

Photomontage Today - Peter Kennard.

Personal photographs

Family photograph albums.

Gallery

Arthur Sikeotis

The Camera

Mark Galer

aims

- To offer an independent resource of technical information.
- To develop knowledge and understanding about camera equipment.

objectives

- Operate the basic controls of a 35mm SLR camera.
- Adjust shutter speed and aperture in response to taking light meter readings.
- Correctly expose a roll of film.
- Care for camera equipment.

Introduction

At first glance the 35mm SLR camera is a sophisticated and confusing piece of equipment. Just as with any other complex piece of machinery the user can, over a period of time, become very familiar with its operations and functions until they are almost second nature. Operating a 35mm SLR camera, just like driving a car, is a skill which is attainable by most people. How long it will take to acquire this skill will very much depend on the individual and how much time they are prepared to spend operating the camera.
Different makes and models of 35mm SLR cameras may appear very different but they all share the same basic features. The features may be placed in different locations on the camera body or operated automatically. If you have any difficulty in finding a particular feature on your own camera ask a teacher or consult your camera manual.

Why a 35mm SLR?

The 35mm SLR is the most popular camera used by keen amateurs and professionals alike. The 35mm format has advantages over larger formats in that a large range of accessories are available and the system is easily portable. The 35mm SLR uses film which is comparatively easy to load. The term 'single lens reflex' describes the way we view the image with this type of camera. The SLR camera has a single lens unlike compact cameras which have two (one for viewing and another for capturing the image onto the film). The single lens of the SLR camera is used to view the subject and take the picture. This is achieved by the use of a mirror behind the lens which reflects the image up into the viewfinder via the '**pentaprism**'. The pentaprism converts the mirror image to one which appears the right way round. If we change the lens, use a coloured filter or change the focus we can see all of these changes through the viewfinder.

Automatic or manual

If your camera operates automatically spend some time finding out how the camera can be switched to manual operation or how you can override the automatic function. Automatic cameras are programmed to make decisions which are not necessarily correct in every situation. A good photographer must be able to use the manual controls of the camera.

Care of the camera

1. Avoid dropping your camera - use a strap to secure it around your neck or wrist.
2. Avoid getting your camera wet - cover the camera when it starts to rain.
3. Only clean your camera lens with a soft brush or a lens cleaning cloth.
4. Never touch the mirror inside the mirror housing or the shutter inside the film back. Both these items are extremely delicate.

Note. Damage to your camera can usually only be repaired by a camera specialist and will usually incur a minimum fee which can be greater than the value of your camera.

Classic style 35mm SLR camera

The basic controls

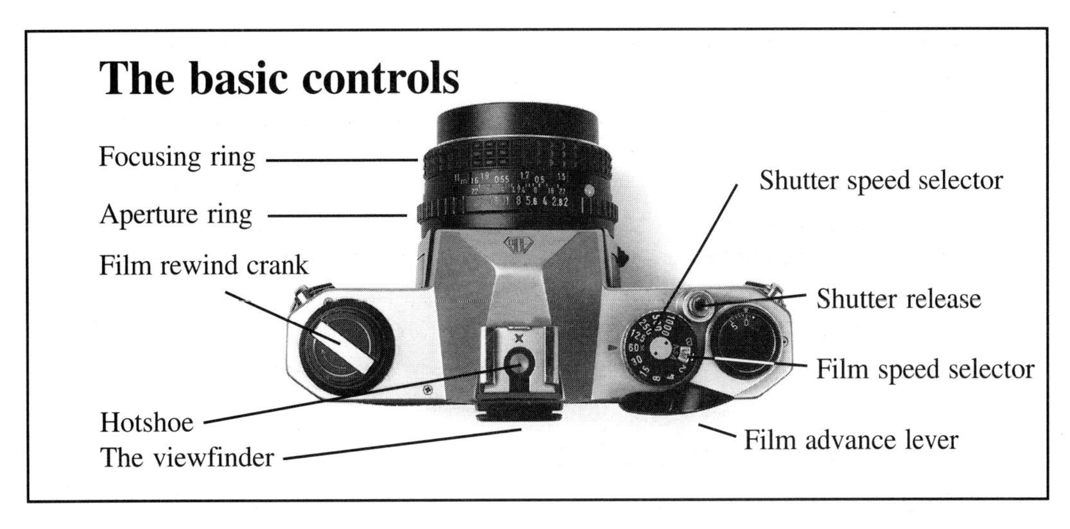

Film back

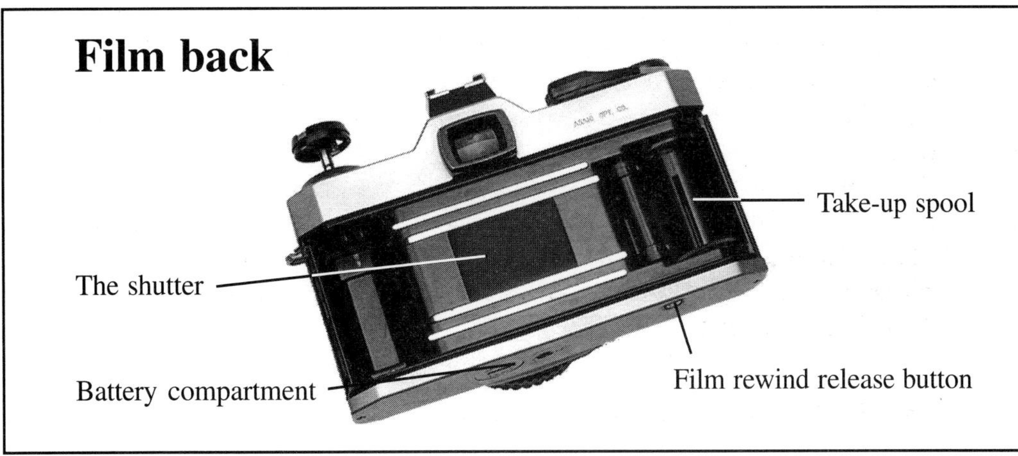

Mirror housing

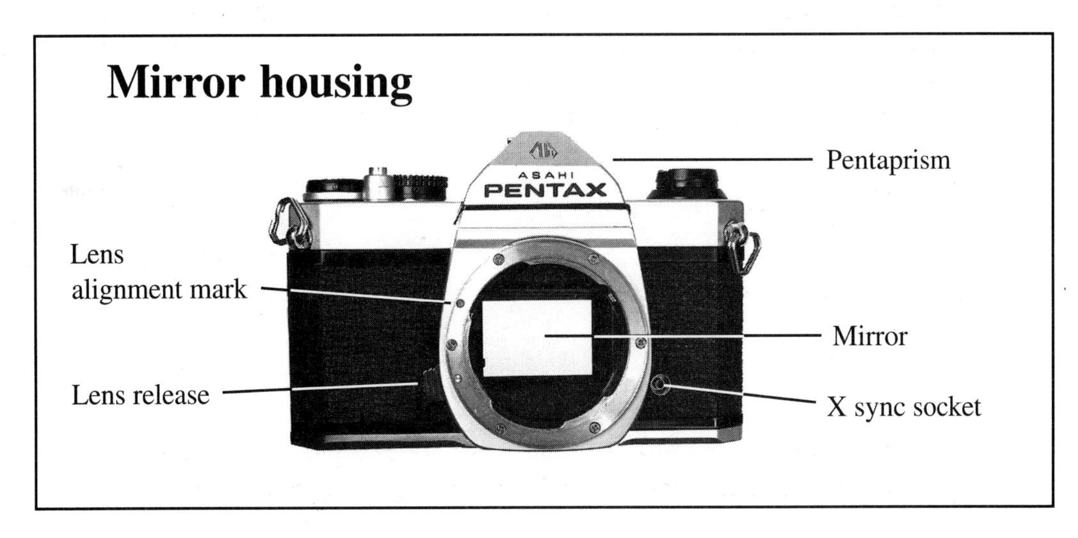

The first film (using a classic style camera)

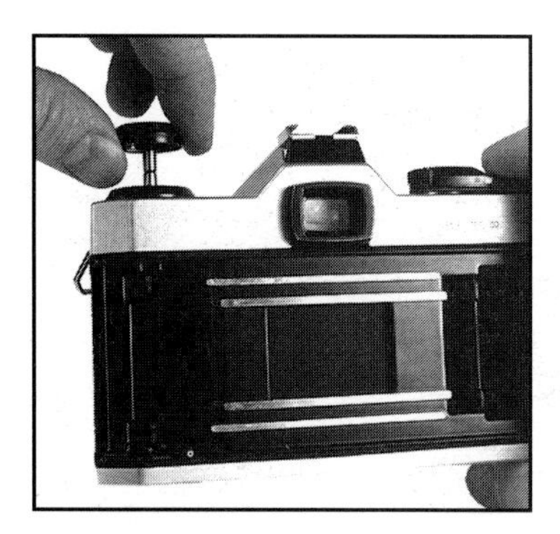

Open the back of the camera. This is usually done by pulling the rewind crank up.

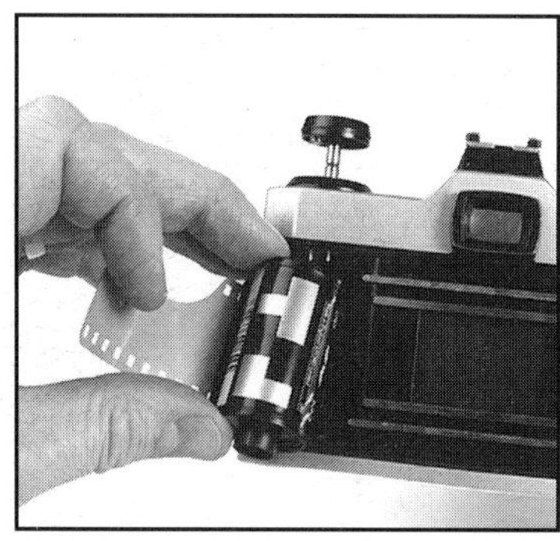

Load the film cassette. Lift the rewind crank to insert the cassette.

Push the rewind crank down and pull a short length of film out of the cassette.

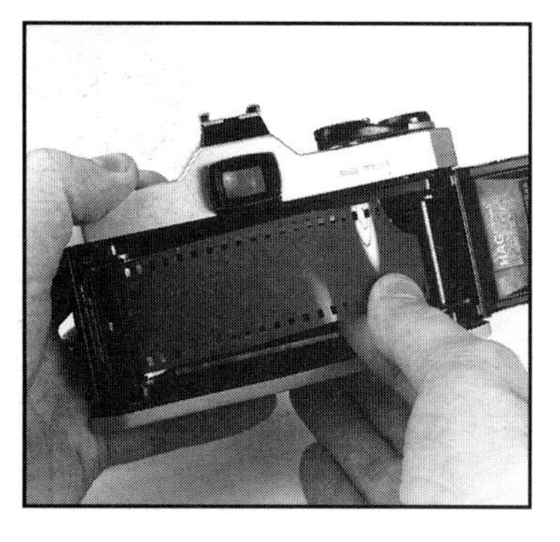

Attach the film leader to the take-up spool. Most of the film leader can be inserted.

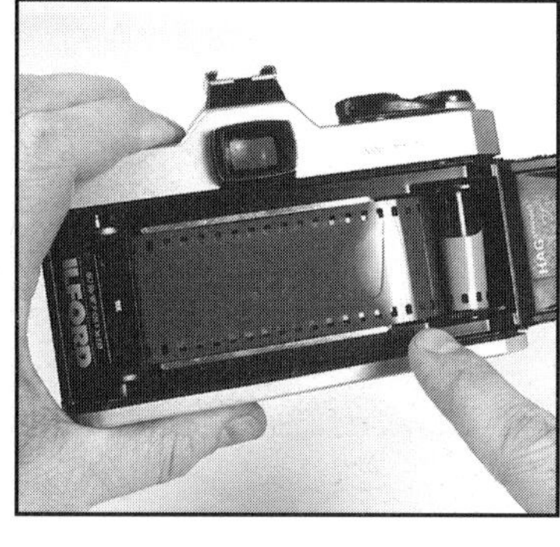

Press the shutter release and advance the film. The teeth must engage with the sprocket holes.

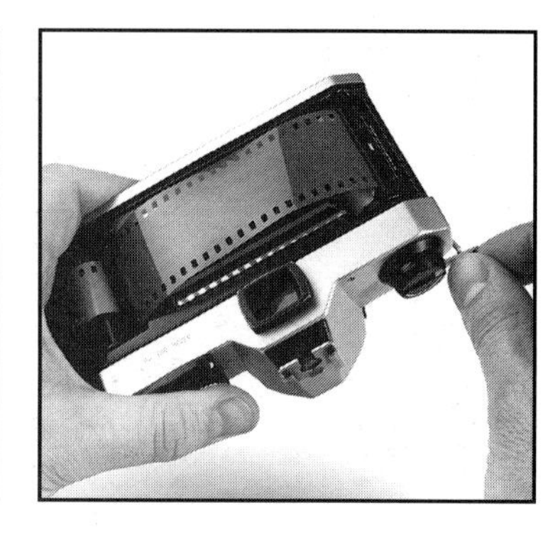

Take up any slack in the film by rewinding the crank handle gently. Close the back securely.

Take two blank shots and advance the film. Check the rewind crank turns each time you advance the film.

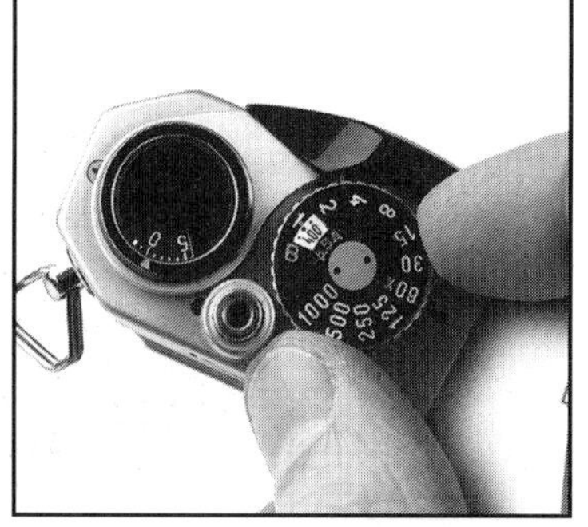

Set the film speed or ISO on the film speed selector. This is not moved again whilst exposing this roll of film.

Select a shutter speed. Usually no slower than 1/60 second unless you are using a tripod.

The first film (continued)

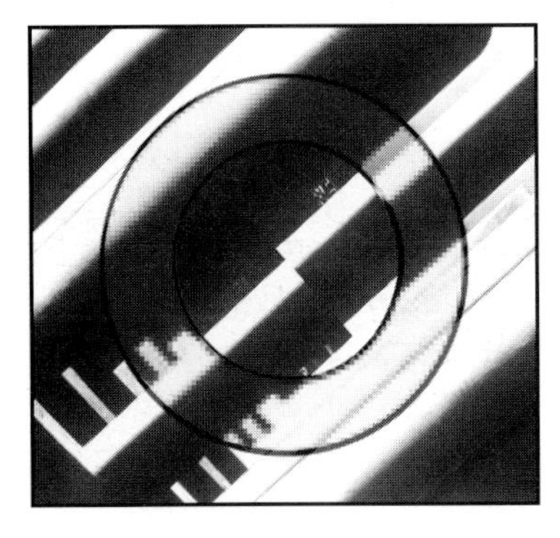

Move the focusing ring back and forth until your subject appears as sharp as possible.

Adjust the aperture and shutter speed until an appropriate exposure is obtained.

Hold the camera firmly, frame your shot and press the shutter release gently.

Advance the film using the film advance lever. The shutter will only fire when fully advanced.

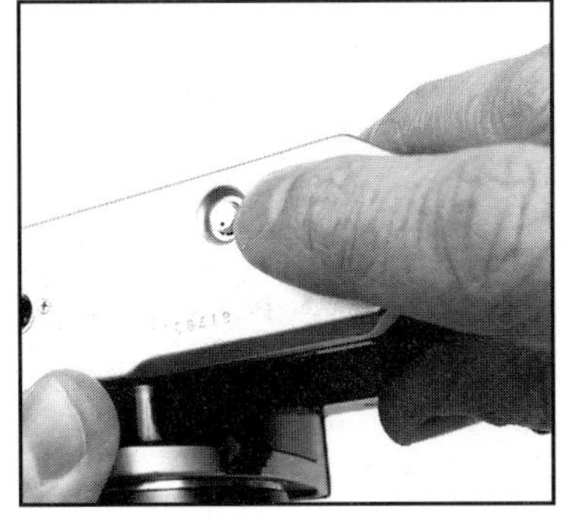

When the last frame has been taken push the film rewind button.

Rewind the film smoothly back into the cassette, approximately one complete turn per frame.

When you feel or hear the film coming away from the take-up spool stop rewinding the film.

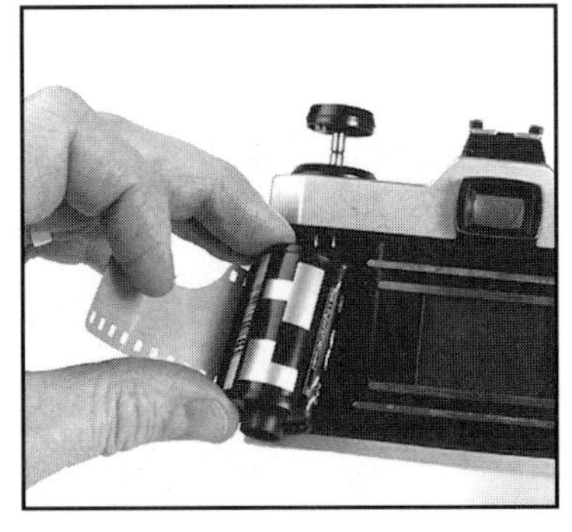

Remove the film by raising the rewind crank. If the film leader is still visible, mark the film to indicate that it is exposed.

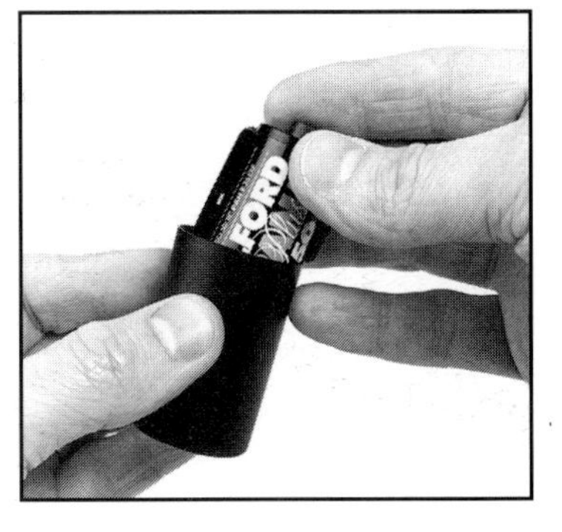

Place the film back into its container until you are ready to process it. This will protect the film from moisture and dirt.

Modern automated cameras

Many modern 35mm SLR cameras include a host of automated features. These include automatic film loading and advance after exposure, focusing, film speed setting and exposure. Exposure compensation and manual exposure should be available if the camera is to be used creatively.

With no film rewind crank, the camera back is usually opened via a lever on the side.

The film leader is aligned to a marker on the camera body before closing the back.

Film speed is set automatically and the exposure mode (auto or manual) is dialled in.

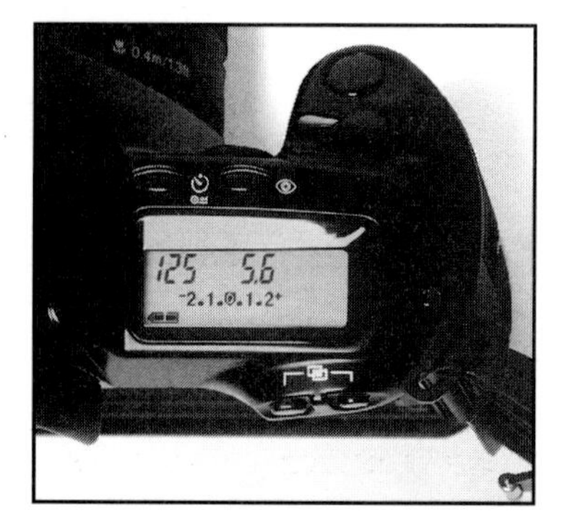

Camera functions and exposure information can be viewed via LCD figures and symbols.

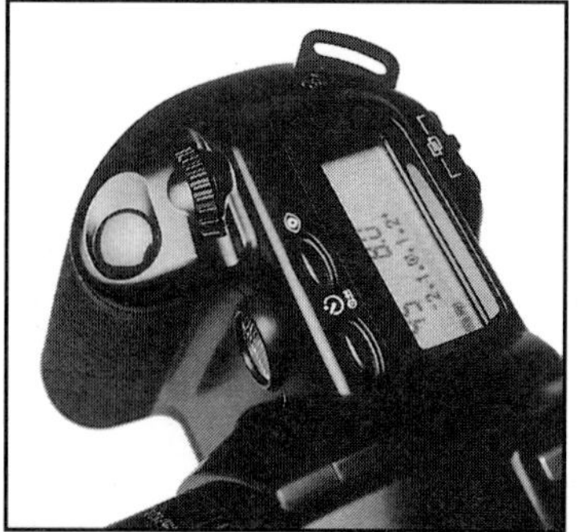

Camera functions are accessed via pressing a function button and rotating a control wheel to select an alternative setting.

Additional controls on some automated cameras allow control over lens apertures, focus and/or exposure lock.

Exposure

The aperture

The aperture of the camera lens opens and closes like the iris of the human eye. Just as the human iris opens up in dim light and closes down in bright light to control the amount of light reaching the retina, the aperture of the camera lens must also be opened and closed in different lighting conditions to control the amount of light that reaches the film. The film requires exactly the right amount of light to create an image. Too much light and the film will be overexposed (negatives appear very dark or dense when processed). Not enough light and the film will be underexposed (negatives appear very light or 'thin' when processed).

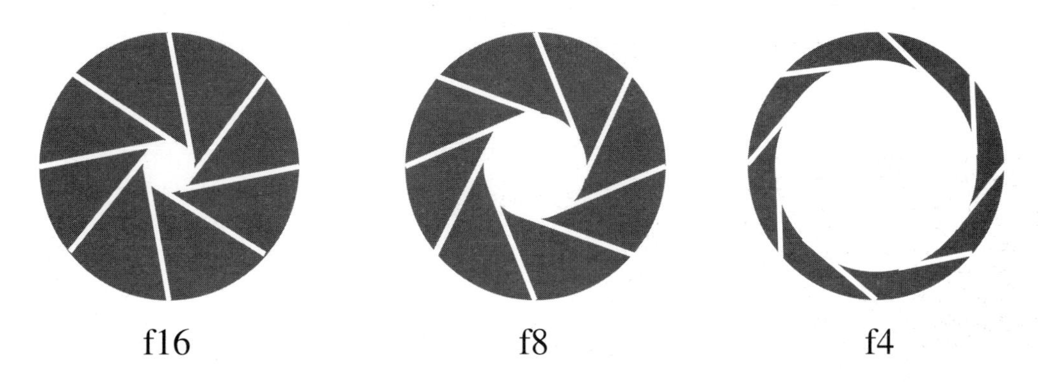

f16 f8 f4

We are guided to select the correct aperture by the light meter. As the aperture ring is turned on the camera lens a series of clicks can be felt. These clicks are called f-stops and are numbered. Depending upon which way you turn the aperture ring every stop lets exactly twice or half as much light reach the film as the previous one. The light meter will advise you when you are getting closer to the correct aperture. The highest f-number corresponds to the smallest aperture and the smallest f-number corresponds to the widest aperture.

The shutter

When the shutter button is pressed the shutter opens for the amount of time set on the shutter speed dial. These figures are in fractions of seconds. The length of time the shutter is open also controls the amount of light that reaches the film, each shutter speed doubling or halving the amount of light. Exposure, therefore, is a combination of aperture and shutter speed.

To slow the shutter speed down is to leave the shutter open for a greater length of time. Shutter speeds slower than 1/60 second can cause movement blur or camera shake unless you hold the camera steady with a tripod or by some other means.

To use a shutter speed faster than 1/250 second requires a wide aperture or a very light sensitive film to compensate for the small amount of light that can pass through a shutter that is open for such a short amount of time. It is suggested that you keep the shutter speed dial on 1/60 or 1/125 second until you are familiar with the equipment that you are using.

TTL light meters

'Through the lens' (TTL) light meters built into cameras measure the level of reflected light prior to exposure. They measure only the reflected light from the subject matter within the framed image. The meter averages out or mixes the differing amounts of reflected light in the framed image and indicates an average level of reflected light. The light meter readings can be translated by the camera's '**central processing unit**' (CPU) and used to set aperture and/or shutter speed.

Centre-weighted and matrix metering

'**Centre-weighted**' and '**matrix metering**', common in many cameras, bias the information collected from the framed area in a variety of ways. Centre-weighted metering takes a greater percentage of the information from the central area of the viewfinder. The reading, no matter how sophisticated, is still an average - indicating one exposure value only. Any single tone recorded by the photographer using a TTL reading will reproduce as a mid-tone, no matter how dark or light the tone or level of illumination. This tone is the midpoint between black and white. If the photographer takes a photograph of a black or white wall and uses the indicated meter reading to set the exposure, the final image produced would show the wall as having a mid-tone.

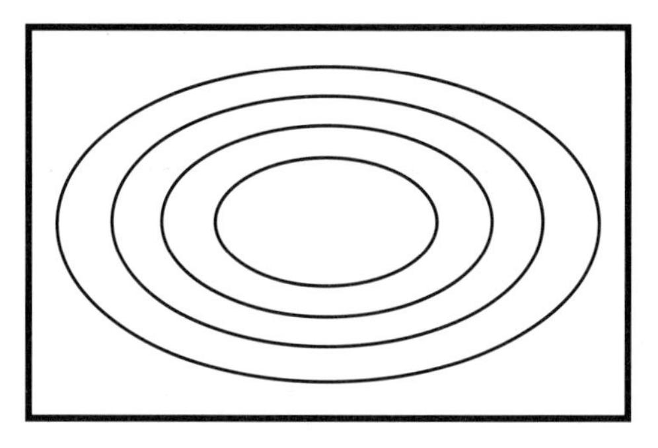

Centre-weighted TTL metering

Automatic TTL exposure modes

If the camera is set to fully automatic or programme mode both the shutter speed and aperture will be adjusted automatically so that the correct exposure will usually be obtained. In low light the photographer using the programme mode should keep an eye on the shutter speed which is being used to achieve the correct exposure. The reason for this is that as the lens aperture reaches its widest setting the programme mode will start to use shutter speeds slower than ones usually recommended to avoid camera shake. Many cameras alert the photographer when this is happening using an audible signal. This should not be treated as a signal to stop taking photographs but as one to take precautions to avoid camera shake such as bracing the camera in some way.

Using the light meter

Most light meters in 35mm SLR cameras will show the light meter reading inside the viewfinder. This allows the photographer to alter the camera settings to achieve the correct exposure without having to remove the camera from the eye.
Inside the viewfinder you should see either a needle, a series of lights or an LED display. Changes should occur when the camera is pointed towards the light. The light meter in a 35mm SLR camera requires a healthy battery to work so if you cannot see any changes consult a teacher or camera dealer.
Most 35mm SLR cameras will have a metering system which will resemble one of the four below. If your metering system is different seek advice.

1. Position the needle between the + and - symbols by altering the aperture or shutter speed. The box containing the + and - symbol may be replaced by a second needle or a circle.

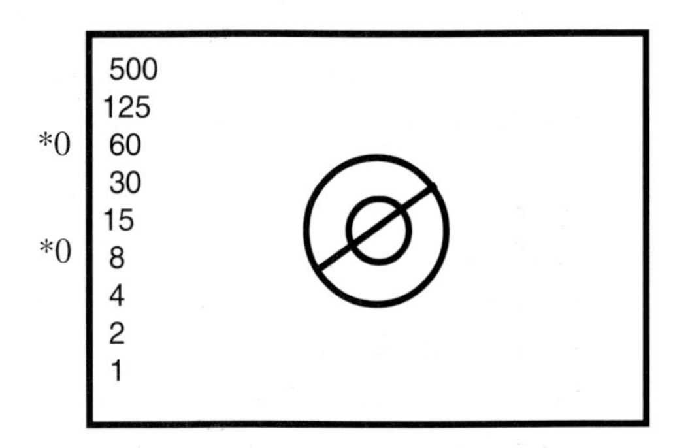

2. One light indicates the shutter speed that has been selected. Move the other light or lights towards this one by altering the aperture or shutter speed until only one light shows.

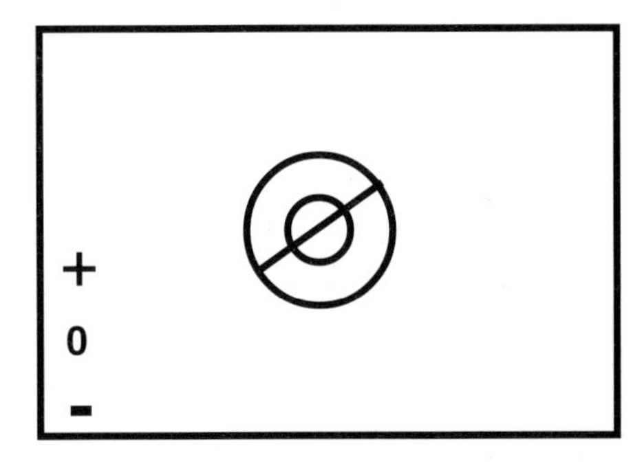

3. Move the aperture or shutter speed until the symbol '0' is illuminated.
Too much light is indicated by a +
Too little light is indicated by a -.

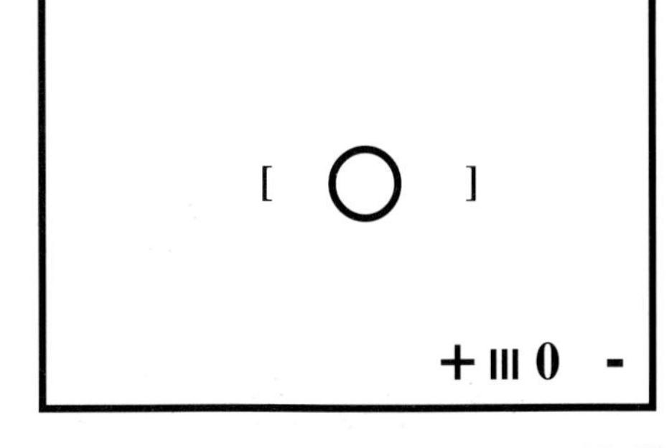

4. Decrease or increase exposure as guided by a series of bars which are displayed (if there is insufficient or excessive exposure) either side of the 0 symbol.

Accurate metering

When taking a light meter reading try to avoid pointing the centre of the viewfinder towards a bright light or tone when setting the exposure. Reposition the centre of the viewfinder on the average tones of your subject instead. Set the exposure and then reframe the image.

Dominant tones

The 'through the lens' (TTL) light meter in a 35mm SLR camera measures the level of reflected light from the subject in the framed image. The TTL meter does not measure the level of illumination from the light source (ambient light). The light meter reading is an average between the reflected light from all the tones framed. When light and dark tones are of equal distribution within the frame this average light reading is suitable for exposing the subject. When light or dark tones dominate in the image area they overly influence the meter's reading and the photographer must increase or decrease exposure accordingly.

When light or bright tones dominate increase the exposure from that indicated.
When dark tones dominate decrease the exposure from that indicated.

Note: Adjustments to the aperture or shutter speed will be compensated for in automatic modes. The adjusted exposure must therefore be made using the exposure compensation function (see 'Light - Exposure compensation').

Aperture priority (Av)

This is a semi-automatic function whereby the photographer chooses the aperture and the camera selects the shutter speed to achieve the indicated exposure. This is the most common semi-automatic function used by photographers.
When aperture priority is selected the photographer needs to be aware of slow shutter speeds in low light conditions or when using small apertures. To avoid camera shake and unintended blur in these instances the photographer has a number of options:

- Select a wider aperture.
- Use a faster film (one with a higher ISO).
- Mount the camera on a tripod.
- Use a lens with a brighter aperture.
- Brace the camera by supporting or resting your elbows, arms or hands on or against a stable surface.

With practice it is possible to use shutter speeds slower than the minimum recommended by the manufacturers.

Shutter priority (Tv)

This is a semi-automatic function whereby the photographer chooses the shutter speed and the camera selects the aperture to achieve the correct exposure. In choosing a fast shutter speed the photographer needs to be aware of underexposure when the light levels start to drop. The fastest shutter speed possible is often limited by the maximum aperture of the lens. In choosing a slow shutter speed the photographer needs to be aware of over-exposure when photographing brightly illuminated subject matter. Movement blur may not be possible when using fast film in bright conditions.

Process & Print

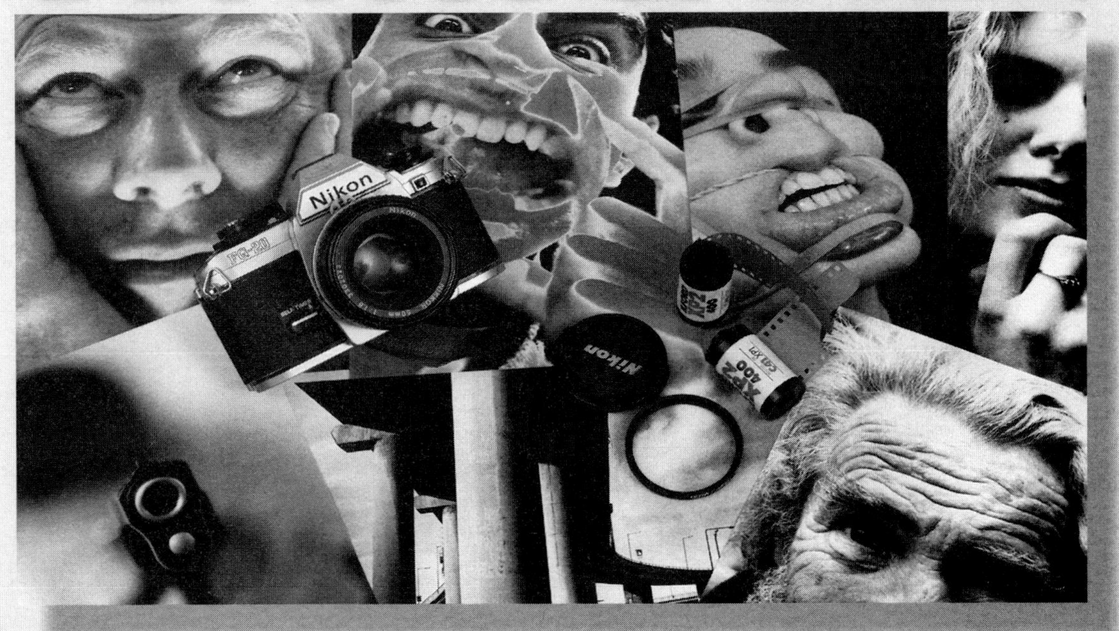

Mark Galer

aims

- To offer an independent resource of technical information.
- To develop knowledge and understanding about the technical processes and procedures involved with producing photographic images.

objectives

- Develop black and white photographic film in daylight developing tanks.
- Produce proof sheets from black and white film for editing purposes.
- Produce clean black and white prints with acceptable sharpness and contrast.

Film processing

You must use a light-tight darkroom and be able to lock the door behind you if possible. You must have with you a DRY film spool, black centre column for the spool, developing tank, tank lid, scissors and film cassette opener if needed.

1. Lights on

If you have rewound your film leaving the leader out, or have managed to retrieve the leader using a special device, the following can be carried out with the lights on.
Cut the leader off between the sprocket holes and feed the first 4 cm onto the spool.

2. Alternative with the lights off

If the leader is wound into the cassette perform the following in total darkness.
Practise loading a blank film onto a spiral in daylight before attempting this in the dark.
Open the cassette in total darkness with a special tool similar to a bottle opener.
Gently withdraw the film from the cassette and remove the film leader with a pair of scissors. Check that no ragged corners are left.
Let the film hang down and push the film gently into the gate of the spiral.

3. Load film in total darkness

Wind the film gently onto the spiral by twisting one side of the spiral back and forth.
If the film jams, remove it from the spiral and start again. Check the spool is dry and the leading edge of the film is not damaged.
When the film is wound onto the spiral remove the film spool (cut or tear the tape).
Push the loaded spiral onto the centre column and then place into the tank. Check that the centre column base is placed down towards the base of the tank.
Place the lid on the tank ensuring that the lid is not twisted in the process.
Check with the other students before switching any lights on.

4. Prepare the chemicals

Fill a 2 litre jug with water at 20°C. Use a thermometer and mix hot and cold water if necessary. If the film developer is very cold or warm this temperature may be raised or lowered slightly to compensate.
Measure accurately with a measuring cylinder the correct amount of film developer for each film to be developed (if the developer is concentrated a small measuring cylinder should be used to gain an accurate reading).
Pour the developer into a measuring cylinder containing enough water to make up the total quantity needed.
Have ready measuring cylinders with the required amount of stop and fix.

Note. It is recommended to prepare enough chemicals to completely fill the developing tank regardless of the number of films being processed.

Film processing (part 1)

Lay out the equipment so that you will be able to find everything in the dark.

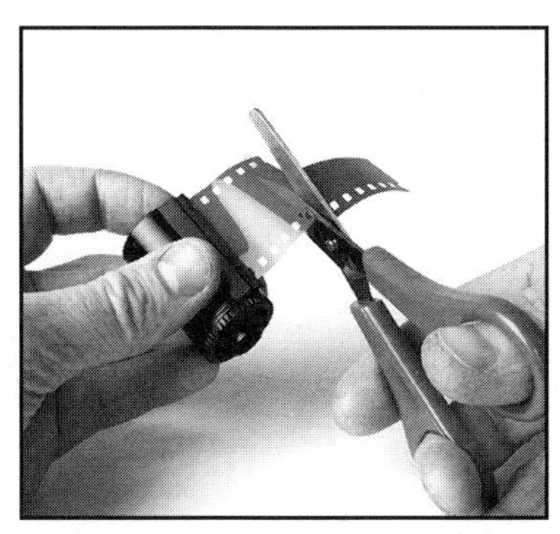

Cut off the leader between the sprocket holes and feed a few centimetres onto the spiral.

Switch off the light and pull the film cassette gently down until about a metre of film is free.

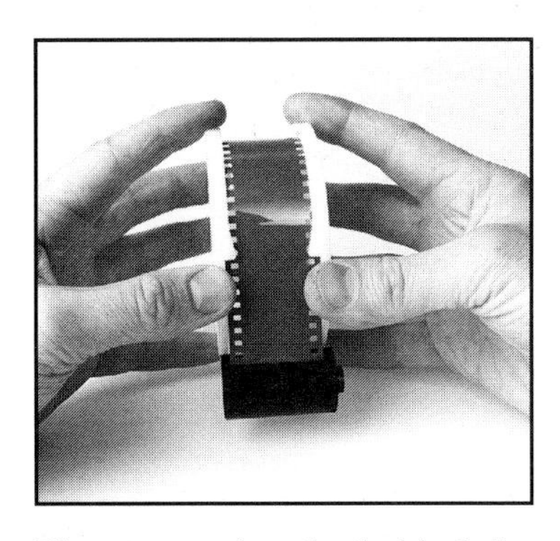

Place your thumbs behind the entrance to the gate as you wind the film onto the spiral.

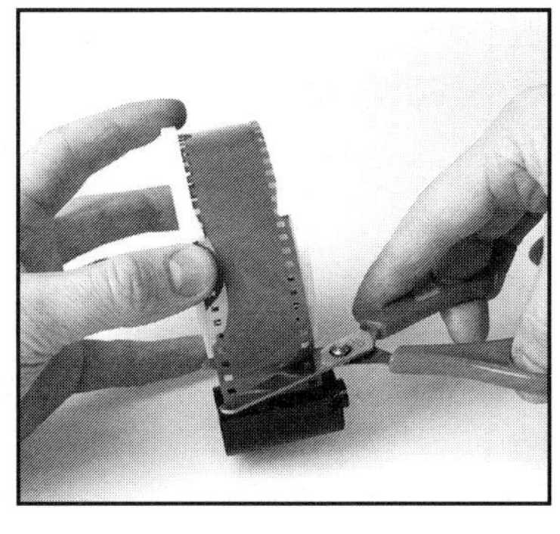

When no more film is left in the cassette, remove it using a pair of scissors.

Place the spiral onto the column and ensure the lip is facing towards the base of the tank.

Turn the lid of the tank clockwise until it locks into place.

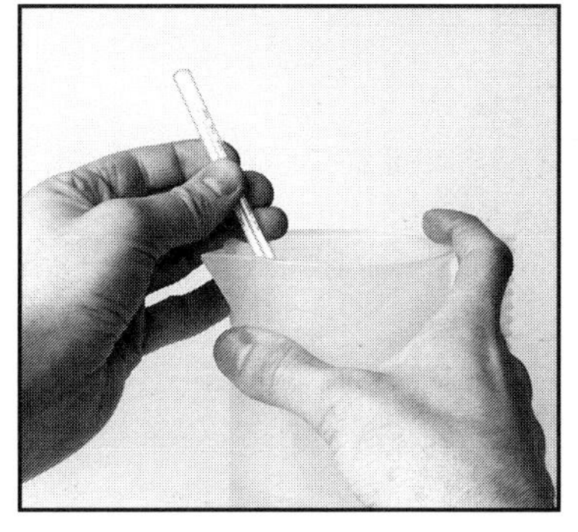

Prepare a jug full of water at 20°C. Measure this accurately with a thermometer and adjust the temperature if necessary.

Measure out the film developer into a small cylinder and make up to the required amount using the water from the jug.

5. Presoak

Fill the developing tank with water from the mixing jug (20°C). Agitate the tank gently for 1 minute to bring the tank and film up to temperature and to soften the film emulsion.

6. Develop

Pour the solution into the tank quickly and start timing immediately.
Press on the liquid seal firmly and invert the tank continuously for 20 seconds.
After every minute give a further 2 inversions of the tank (5 seconds agitation in total).
Start to pour the developer away 15 seconds before the developing time has elapsed.

7. Stop

Fill the tank with STOP BATH from the measuring cylinder.
Replace the cover and invert for 30 seconds.
Return the stop bath to the correct container using a funnel (if stop turns blue, pour away).

8. Fix

Fill the tank with fixer from the measuring cylinder which has already been diluted.
Replace the cover and invert once every 15 seconds for 5 minutes.
Return the fixer to the correct container using a funnel.

9. Wash

Fill the tank with water from the mixing jug, replace the seal and invert several times.
Pour the water away and remove the tank lid.
Check that the film is clear (no milkiness) and wash under running water for 15-20 minutes.

10. Final rinse

Remove the films from the tank and add a few drops of wetting agent to the water.
Replace the films and waterproof seal, invert twice and leave for 30 seconds.

11. Dry

Remove the films from the tank and remove from the film spiral.
Shake the excess water from the spiral (a flicking of the wrist removes most of the water).
Hang the films up to dry with a weight on the bottom so that they dry straight.
Hang the films to dry in a dust-free place.
Do not dry the film with a hair-dryer as this may damage the film.
Film should dry in 10 minutes in a film dryer or 2 hours if hung on a line.
Do not use excessive heat to dry your films.

12. Clear away

Replace all the lids on the chemical bottles.
Rinse all parts of the tank and mixing equipment and leave to drain.

Film processing (part 2)

Pour in the developer quickly, start timing and attach the liquid seal to the tank.

Invert the tank for 20 seconds and then twice every minute.

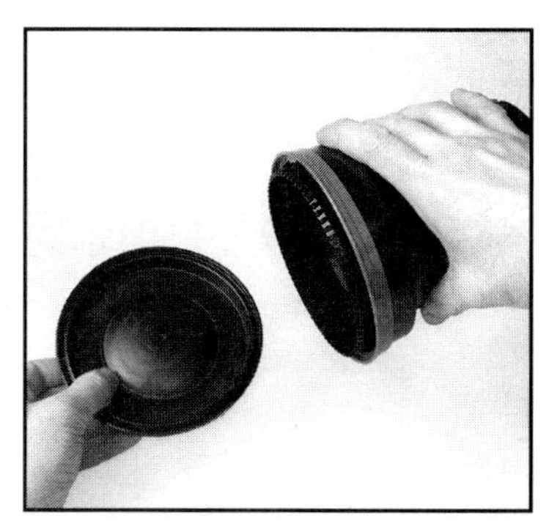

After the correct time has elapsed pour the developer away.

Fill the tank with stop bath, replace the cover and invert for 15 seconds.

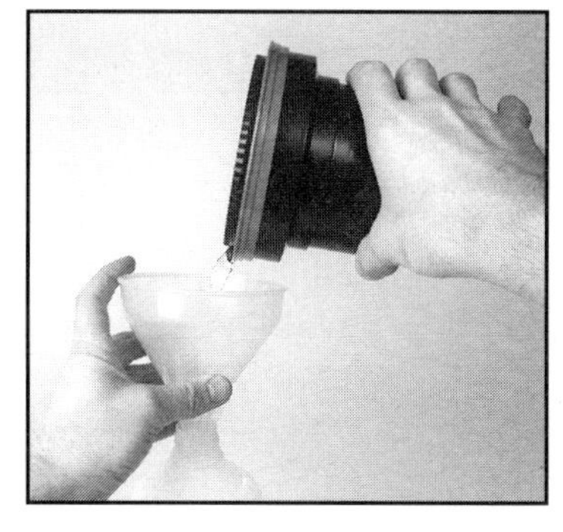

Pour the stop bath back into its container and then fill the tank with fixer.

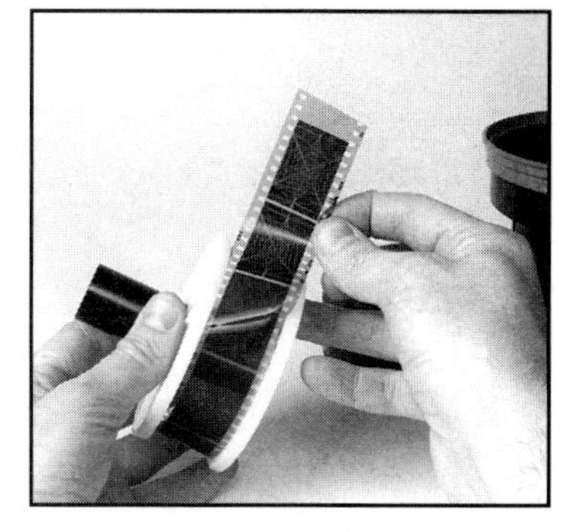

Save the fix, rinse the tank out with water and then inspect the film to ensure that it is clear.

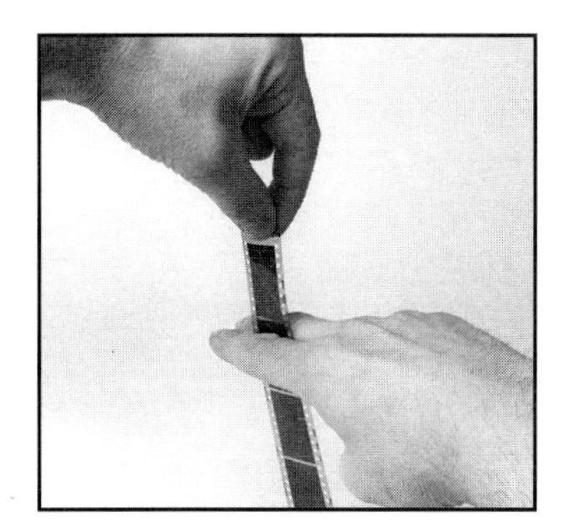

Wash the film for 20 minutes. Add a few drops of wetting agent to the final wash. Shake all water from the spiral.

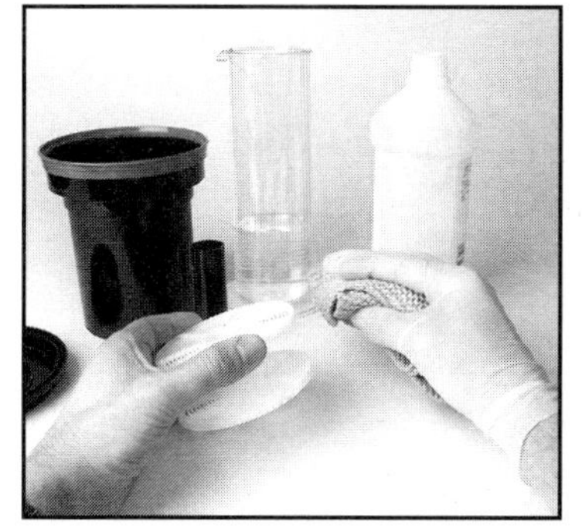

Hang the film up to dry in a dust-free area with a weight attached to the base. Wash all the equipment and surfaces.

Replace all of the tops on the chemical containers and put the tank and spiral in a warm place to dry.

Print processing

Producing your own prints is a rewarding photographic activity. To get the best results and avoid wasting materials you should adopt a methodical and patient approach.

Health and safety

It is important to protect yourself and your clothes from chemical contamination when processing photographic paper. Although the chemicals are not highly corrosive they can lead to rashes and stained clothes. The following precautions must be taken.

a) Wear a laboratory coat or old shirt at all times whilst processing.
b) When pouring chemicals wear protective gloves and eye protection.
c) Never handle prints with your hands. Use the print tongs provided.
d) Wash contaminated skin or clothes immediately with soap and water.
e) Never sit on work surfaces that have come into contact with chemicals.
f) Never bring food or drink into an area where chemicals are being used.
g) Ensure that the room you are working in is well ventilated.

Processing resin coated (RC) paper

Note: Procedures 1-4 must be carried out in 'safelight' conditions only.

1. Place the exposed paper into a tray of developer at room temperature (approximately 20½C) and submerge it using a pair of print tongs.
2. Rock the tray gently for 1 minute. Do not 'snatch' the paper from the developer if you think it is going too dark. This will only lead to a poor quality print.
3. Use print tongs to transfer the paper to the tray containing stop bath (allow a few seconds for the surplus developer to drip back into the developer tray). The tongs from the developer tray must not touch the stop bath. If they do, wash them before using again.
4. Transfer the paper to the fix after 10 seconds. Fix for 2 minutes rocking the tray gently.
5 Wash the print in running water for 2 minutes. Avoid adding prints to the wash from the fix during this process. Test strips may be examined in daylight conditions after only 15 seconds in the fix and 5 seconds in the wash if they are to be thrown away.
6. Squeegee the water off both surfaces of the print applying light pressure only.
7. Dry the prints face-up on a dry work surface. Prints must not overlap during drying as they will stick together. Drying prints with excessive heat will cause the prints to curl.

Cleanliness

Dirty marks, hairs or white spots appear due to chemical contamination or dirty negatives. Make sure your hands and work-bench are clean and dry.
Use a blower brush to clean negatives.
HANDLE NEGATIVES WITH EXTREME CARE.

Test strips

Test strips are used to measure the brightness of the light coming from the enlarger lens. They are used for the same reason that we take a light meter reading in the camera, i.e. to find the correct aperture and the length of time needed for a correct exposure. Because photographic paper requires more exposure than film we use a timer connected to the enlarger to measure seconds, instead of fractions of seconds. It is, however, possible to count accurately or use the second hand of a clock to measure exposures for photographic paper. There is the temptation to guess an exposure from the first test strip even if all the exposures shown on the test strip are far from correct. If used often, test strips will save you time and photographic paper.

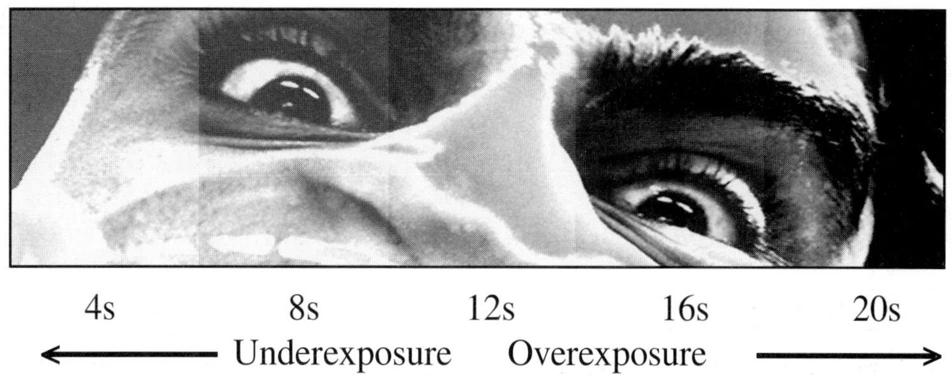

4s 8s 12s 16s 20s

← Underexposure Overexposure →

A good test strip will cross many areas of density and show over- and underexposure.

Contrast control

The term contrast refers to the different tones of grey we can see in a print. A low contrast print is one that may have many shades of grey but no deep blacks or bright whites. A high contrast print is one that has many areas of black and white but few, if any, shades of grey. The contrast can be altered on multigrade paper by using coloured filters.

Low contrast

High contrast

Increasing the Magenta filter on a colour enlarger will increase the contrast on multigrade paper. Increasing the Yellow filter will lower the contrast. Increasing the filtration on an enlarger will require longer exposures to obtain a print of the same density.
Increasing to a higher number on a multigrade head will increase the contrast.
Decreasing the number on the multigrade head will decrease the contrast.

Note. The higher the number of the filter, the higher the contrast.

Making a proof sheet

Preparing your negatives

1. Cut your negatives into strips of five or six frames on a clean dry surface. Do not handle the image area of the negatives with your fingers.
2. Slide the negative strips into the negative file shiny side up (emulsion side down). Store your negatives away from wet or damp work surfaces.
3. Choose an enlarger and switch off any white lights.
4. Switch on the enlarger and adjust the lens aperture so that the light is at its brightest. Close the aperture by 2 stops.
5. Adjust the enlarger head height until the spread of light covers the negative bag or contact printing frame. Switch off the enlarger.

Making a test strip

6. In '**safelight**' only, cut a 2cm strip off a sheet of photographic paper using a pair of scissors or a guillotine. Return the rest of the paper to the bag and seal it.
7. Place the strip of photographic paper, emulsion side up, on the baseboard (on your pearl or gloss paper the emulsion side will reflect the red darkroom lights).
8. Place a strip of negatives of average exposures on top of the strip of paper. If you have a clear negative file you do not need to remove them.
9. Cover the photographic paper and the negative with the clean sheet of glass and ensure that both are in close contact. Press down if necessary.
10. Cover up 3/4 of the test strip with a piece of card.
11. Set the enlarger timer to 4 seconds and expose the first 1/4 of the strip.
12. Uncover one further 1/4 and expose for a further 4 seconds. Repeat process until all the paper has been exposed.
13. Process the test strip in trays of developer, stop and fix.
14. Examine the test strip in daylight conditions. Choose the best exposure from the test strip and set the enlarger timer accordingly. If all the exposures are too light then increase the exposure times for a second test strip. If all the exposures are too dark close the lens aperture down 1 more stop.
15. Expose all of the negative strips onto one sheet of printing paper and process.

REMEMBER: THE BRIGHTER THE LIGHT, THE DARKER THE PRINT.

Common faults

1. Faint and fuzzy image .... Photographic paper upside down.
2. Test strip too dark .... Enlarger lens not stopped down.
3. Test strip too light .... Enlarger lens stopped down too much.
4. Out of focus images .... Glass not in close contact.
5. Marks or splodges .... Glass not clean or chemical contamination.

Making a proof sheet

Cut negatives. Be careful not to drag negatives along the work surface.

Slide negatives into the file sheet. Handle negatives carefully by the edges only.

Stop lens down 2 stops. This will ensure that the exposure time is of a reasonable length.

The spread of light from the enlarger lens must cover the baseboard.

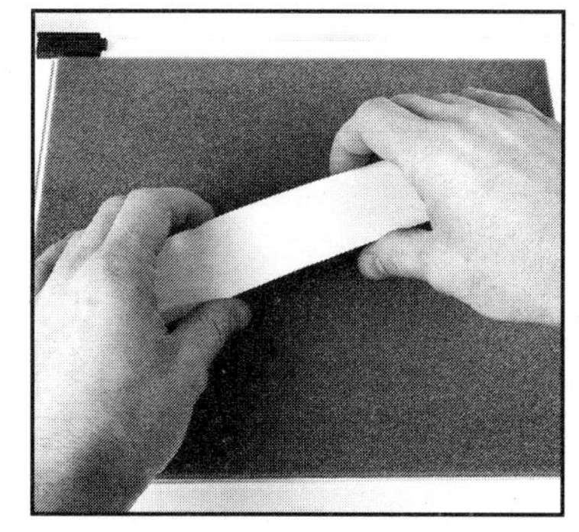

Place the test strip, emulsion side up, on the baseboard of the enlarger.

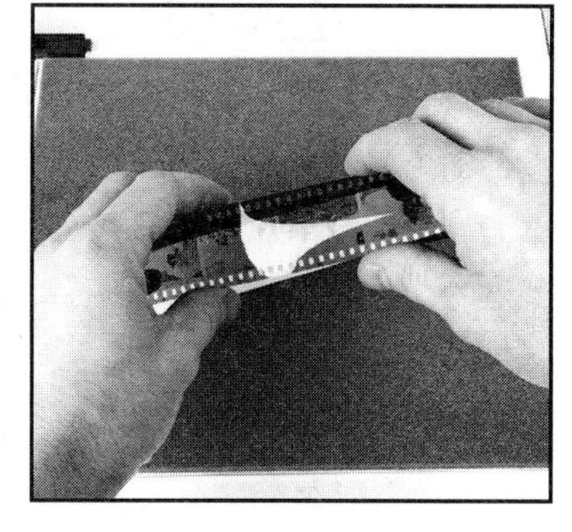

Place the negative onto the test strip, emulsion side down, shiny side up.

Place a clean sheet of glass over the negative and photographic paper.

Mask off a portion of the test strip using a piece of thick card to control exposure.

Process the photographic paper.

Printing

1. Select a negative to print from the contact sheet.
2. Remove the negative carrier from the enlarger. Ensure that it is switched off.
3. Place the negative in the carrier, emulsion side down, glossy side up.
4. Blow any dust off the negative using a blower brush.
5. Replace the negative carrier in the enlarger.
6. Place printing easel on the enlarger baseboard and switch on the enlarger.
7. Ensure the aperture is at its largest f-stop to supply maximum light for focusing.
8. Raise or lower the enlarger head to obtain the desired degree of image enlargement.
9. Focus the image (use focusing aid for precise focusing).
10. Compose and frame the picture.
11. Reduce the aperture of the enlarger lens by 2 or 3 f-stops.
12. Turn the enlarger off and place a test strip on the easel, emulsion side up.
13. Expose the test strip at 4 second intervals.
14. Process test strip.
15. Select the best exposure in daylight.

If the exposure time is less than 7 seconds, close the lens down 1 more f-stop. If it is greater than 30 seconds open up the lens 1 f-stop.
EACH F-STOP WILL DOUBLE OR HALVE THE EXPOSURE TIME.

Burning-in

When a section of the print appears too light it is possible to increase the exposure to this area without affecting the rest of the print. This is called '**burning-in**'.
To burn in use a large piece of card with a hole cut in it. The shape of the hole can vary depending upon the area to be burnt in, but it should be much smaller than the area needing the extra exposure. Using test strips, find out the additional exposure needed for the area to be burnt in. Expose the entire paper for the shorter exposure and then place the card with the hole approximately midway between the lens and the paper for the additional exposure. The light falling through the hole should be directed onto the area needing the additional exposure and the card should be kept moving during this exposure so that no hard edge between the original and extra exposure appears on the final print.

Holding back or dodging

When a small area of the print appears too dark it is possible to reduce the exposure to this area only. This is called '**dodging**'. The selected area is shaded for part of the exposure by placing a small piece of card taped to a piece of wire held between the enlarger lens and photographic paper. The amount of time the area needs to be shaded can be determined using a test strip, and as in the burning-in process the card should be placed midway between the enlarger lens and the photographic paper and kept moving.

Printing

Place the negative into the carrier, shiny side up, emulsion side down.

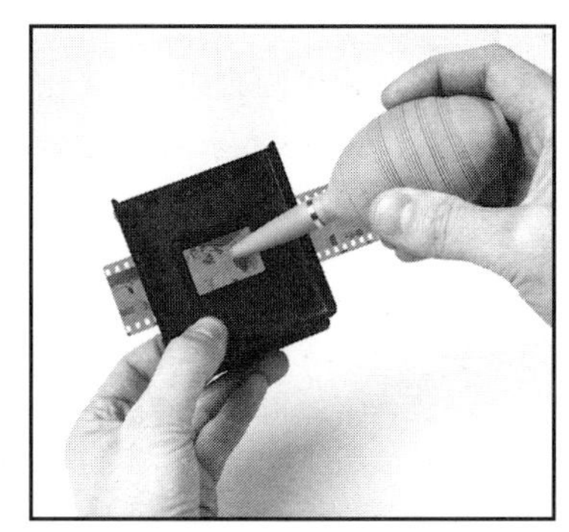

Remove any dust from the negative using a blower brush or soft cloth.

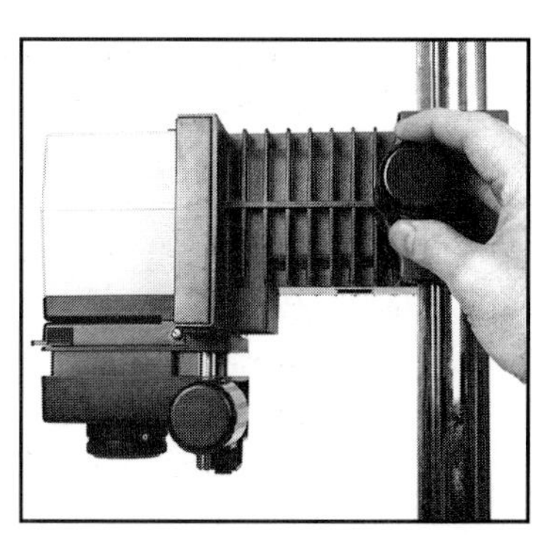

Adjust the height of the enlarger head to create an image of the right size on your printing easel.

Focus the image by eye with the enlarger lens set to its maximum aperture.

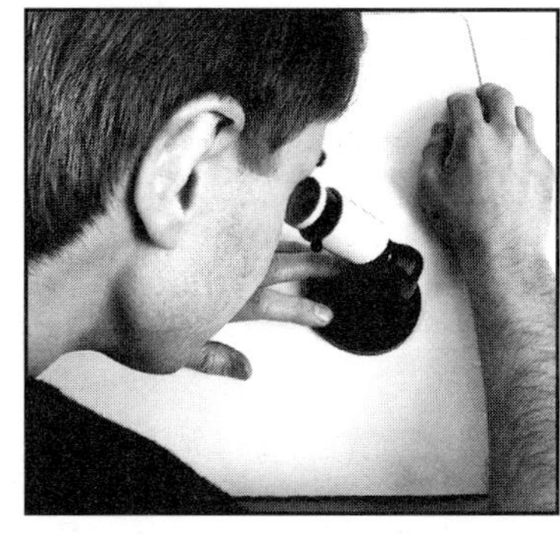

Use a focus finder to ensure accurate focusing.

Reduce the aperture by 2 stops to ensure a reasonable length of exposure and sharp focus.

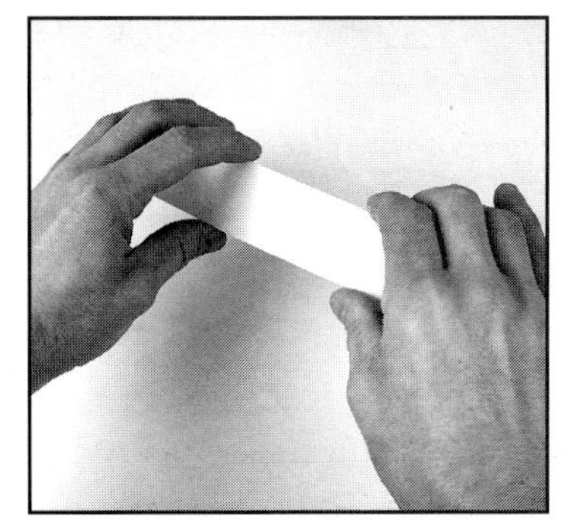

Expose the test strip from a section of the projected image which has a variety of densities.

Process the test strip. Ensure that the test strip is processed for the full processing time.

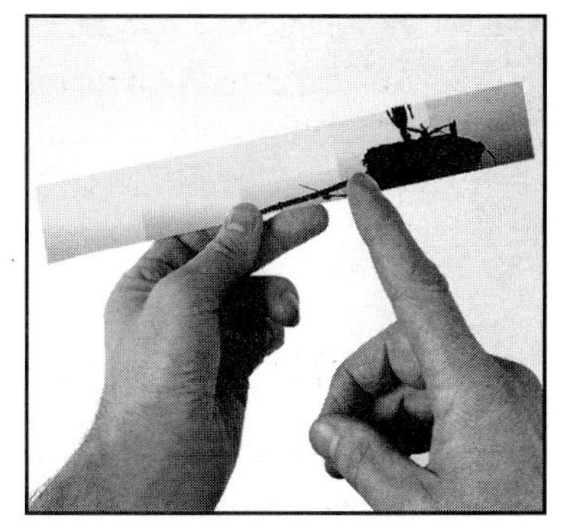

View the test strip to gauge the correct exposure and contrast before making the final print.

Quality control of negatives

High quality photographic prints are made from high quality negatives. If the negatives have not been exposed or processed correctly it is difficult to obtain high quality prints. Negatives should be examined carefully after processing each film to ensure that quality has been achieved or maintained. Negatives should be examined using a '**light box**'.

Exposure

Correct exposure is determined by looking to see how the shadows and dark tones of the subject have recorded onto the film (these are the lighter/clearer areas of the negatives). If these areas are clear with no visible detail then the negatives are underexposed. This underexposure may be caused by either:

- Failing to compensate for high contrast subjects or large areas of bright light in the framed image (such as the sky or a window).
  Improve compensation techniques (see '**Light - Exposure compensation**').
- The use of an inaccurate light meter.
  Alter the ISO (film speed) to correct an inaccurate light meter
  Example 1: Rate a 400ISO film at 200ISO to increase the exposure by 1 stop.
  Example 2: Rate a 400ISO film at 800ISO to decrease the exposure by 1 stop.

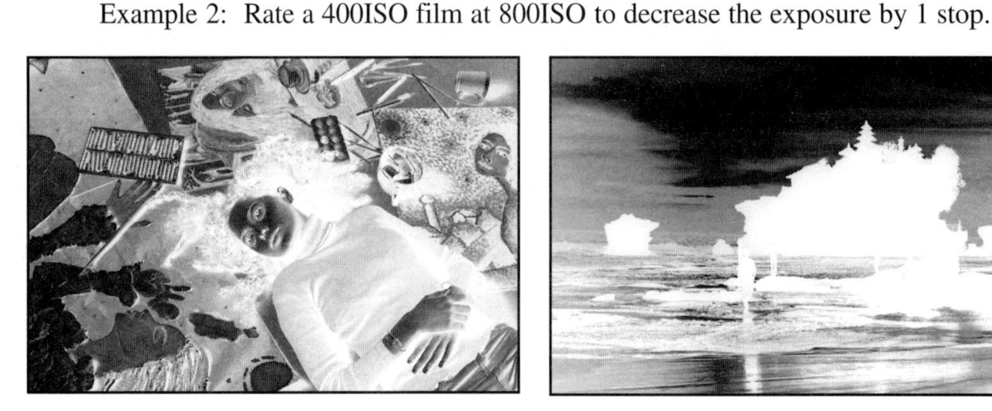

Correctly exposed negative *Underexposed negative (no detail in shadows)*

Processing

Correct exposure is determined by looking to see how the light tones of the subject have recorded (the darker areas of the negative). The film information (film name and frame numbers) can also be examined. If these tones are pale and the contrast of the negatives are low then under-processing is likely.
If the light tones of the subject are very dark and film edge information appears bloated and woolly then over-processing is likely.
Reduce or increase processing time or temperature to change processing problems.

digital

al imaging digital imaging digital imaging digital imaging digital ima

Digital

Capture and Output

Pixellated Eye

aims

- ~ To offer an independent resource of technical information.
- ~ To develop knowledge and understanding about the technical processes and procedures involved with producing a digital print.

objectives

- ~ **Create, capture and output** a digital image file using knowledge concerning:
  - image mode
  - file format and size
  - resolution

Introduction

The term '**digital photography**' is used to describe images that have been captured by digital cameras or existing photographs that have been scanned to create digital image data. The term also describes the processing of digital image data on computers and the output of '**hard copies**' or digital prints (on paper or plastic) from this data.
Digital photography is now revolutionising not only the process of photography but also the way we view photography as a visual communications medium. This new photographic medium affords the individual greater scope for creative expression, image enhancement and manipulation.

Digital foundations

This guide is intended only to lay the foundations of practical digital knowledge. The individual may find it beneficial to supplement this information with additional guides specific to the equipment and computer programs being used. The information supplied by these additional guides, although valuable, may quickly become redundant as new equipment and computer programs are released frequently in this period of digital evolution.

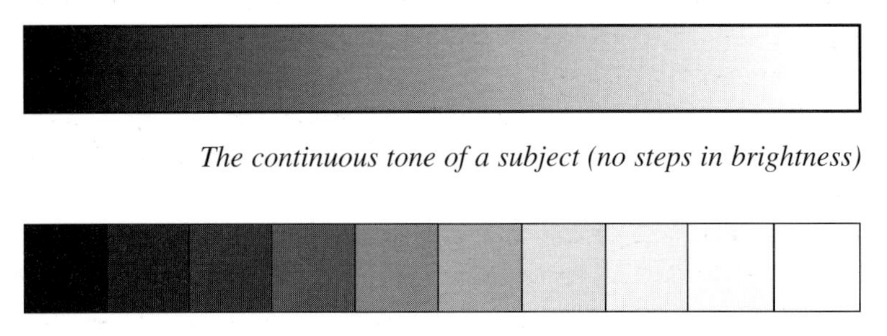

The continuous tone of a subject (no steps in brightness)

Ten pixels each with a different tone or level used to describe the above

Pixels and levels

A digital image is one in which the image is constructed from '**pixels**' (picture elements) instead of silver halide grains. Pixels are square and positioned in rows horizontally and vertically to form a grid. Each pixel in the grid is the same size and is uniform in colour and brightness, i.e. the tone does not vary from one side of the pixel to the other.
In the illustration above 10 pixels, each with a different tone, are used to describe the '**continuous tone**' above it. Each different tone is called a '**level**' and assigned a numerical value, e.g. 0 to 9.
In a typical digital image there are 256 different levels or separate tones to create a smooth transition from dark to light. If the pixels are sufficiently small when printed out the viewer of the image cannot see the steps in tone thereby giving the illusion of continuous tone.

Modes and channels

There are a number of different models or formats used for describing the tone and colour information in a digital image. These models are called '**modes**', e.g. a black and white image can be captured in '**bitmap**' mode or '**greyscale**' mode. In a bitmap image each pixel within the grid is either black or white (no shades of grey). This mode is suitable for scanning line drawings or text. For subjects with continuous tone the greyscale mode is used.

Greyscale

Black and white (continuous tone) photographs are captured or scanned in what is called 'greyscale'. Each pixel of a greyscale image is assigned one of 256 tones or levels from black to white. These 256 levels allow a smooth gradation between light and shade simulating the continuous tone that is achieved with conventional silver-based photography. A greyscale image is sometimes referred to as an '**8-bit image**' (see '**File size**').

RGB

A '**full colour**' image can be assembled from the three primary colours of light; red, green and blue or **'RGB'**. All the colours of the visible spectrum can be created by varying the relative amounts of red, green and blue visible . The information for each of the three primary colours in the RGB image is separated into '**channels**'. Each channel in an RGB image is usually divided into 256 levels. An RGB colour image with 256 levels per channel has the ability to assign any one of 16.7 million different colours to a single pixel (256 x 256 x 256).

Colour images are usually captured or scanned in the RGB 'colour mode' and these colours are the same colours used to view the images on a computer monitor. A colour pixel can be described by the levels of red, green and blue, e.g. a red pixel may have values of Red 255, Green 0 and Blue 0; a yellow pixel may have values of Red 255, Green 255 and Blue 0 (mixing red and green light creates yellow); and a grey pixel may have values of Red 128, Green 128 and Blue 128.

Note. RGB is only one of several colour modes. 'CMYK' is a colour mode that is used primarily in the printing industry and uses the colours cyan, magenta, yellow and black. RGB images should only be converted to CMYK by experienced digital operators.

Describing colour

Hue, saturation and brightness are terms used to describe the three fundamental characteristics of colour. '**Hue**' is the name of the colour, e.g. red, orange or blue. '**Saturation**' is the purity or strength of a colour, e.g. if red is mixed with grey it is less saturated. '**Brightness**' is the relative lightness or darkness of the colour.

Capture

A digital image can be captured directly from the subject using a digital camera or scanned from an existing photographic image. The digital information acquired by the camera or scanning device can be read by a computer. It can then be enhanced, manipulated, stored, output to the web or printed.
We can view the digital process as a chain of events or processes:

- Digital capture, creation and storage.
- Digital image data processing (manipulation and enhancement).
- Digital image output (the creation of hard copies or prints).

Digital cameras

Digital cameras use a light sensitive **'CCD'** (charge coupled device) instead of light sensitive film to capture an image. They have been adapted from their use in video cameras and flatbed scanners. The advantage of a digital camera is that the CCD, unlike film, is reusable. It is able to 'download' (transfer) its information to a computer disk after every captured image. The image is usually able to be viewed as soon as it is captured (without the need for chemical processing) and the memory of this image can be stored on a computer disk.
The negative aspects of digital capture are the comparative loss of quality when compared to the film version and the relative high cost of digital cameras.

Digital image capture devices

Scanners

Digital images can be created from existing photographic images. Prints can be scanned on flatbed scanners or the information can be taken directly from the film using a film scanner. Scanners, especially flatbed scanners, are now very affordable and often come with **'software'** (a computer program) with which to manipulate or prepare the image for digital printing.
It is possible to scan film with some success on some high-end flatbed scanners but dedicated film scanners are required for optimum quality.

Making a comparison

Digital images can be captured either by using a digital camera or from a film or print that has been scanned. The decision between the two methods of digital capture can be a difficult one to make as there are benefits to each method. It is a decision that is facing professionals and amateurs alike. In making a decision as to whether it is advantageous to purchase a digital camera the following points should be considered:

~ Creative control
~ Quality
~ Cost
~ Convenience

Creative control

Establish whether the digital camera offers the individual control over the lens aperture and the duration of the exposure (the creative controls for depth of field and movement blur).

Quality of image

What is the comparable quality between images captured with digital cameras and images scanned from film or photographic prints?
With small enlargements the limiting factor is usually the printer being used and no difference between the two will probably be noticeable.
At larger print sizes the scanned image from film may look superior to the image captured by a digital camera. The limiting factor to the quality may now be the CCD of the digital camera. Even digital cameras with '**megapixels**' cannot capture as much detail as conventional film. A digital camera with a CCD capable of capturing at least two million pixels is required to produce reasonable quality A4 prints (see '**Resolution**').

Cost

What is the cost of a digital camera (including the additional equipment required to transfer the images to the computer) compared with a conventional 35mm SLR (including the cost of processing equipment, film scanners and/or darkrooms)?
Capturing images using digital cameras is cheaper, frame for frame, when compared to film. This takes into account the cost and effective life of the batteries required to run a digital camera and the safe handling and disposal of chemicals (health and safety).

Convenience

The method and speed of transferring images from a digital camera to the computer must be compared to the time spent processing and scanning film. Some photographic labs are able to offer a scanning service.

File size and format

Digital images can take up large amounts of computer memory. The simple binary language of computers and the visual complexities of a photographic image lead to large '**file sizes**'. The written contents of a book in the form of a text document may comfortably fit onto a floppy disk whereas only a small portion of the cover image may be stored on the same floppy disk.

File sizes

The binary digit or '**bit**' is the basis of the computer's language. One bit is capable of two instructions and can describe a pixel in two tones (black or white).

Two bits can give four instructions and 8 bits or '**byte**' can give 256 instructions (the number required to describe the tonal value of each pixel in a greyscale image). 1024 bytes is a '**kilobyte**' (Kb) and 1024 kilobytes is a '**Megabyte**' (Mb). A floppy disk is capable of storing just over 1 megabyte (1Mb). The '**digital file**' of the image that is used on the cover of a glossy magazine is likely to exceed 20 megabytes (20Mb).

Fortunately files this large can be '**compressed**' (reduced in memory size) for storage and it is possible to fit a large image file onto a floppy disk. Storing large files on floppy disks, although possible, is not practical. Removable hard drives (such as the inexpensive '**Zip**' drive made by Iomega) are commonly used for storing and transferring large image files conveniently and quickly.

File formats

When an image is captured by a camera or scanning device it has to be '**saved**' or memorised in a '**file format**'. If the binary information is seen as the communication, the file format can be likened to the language or vehicle for this communication. The information can only be read and understood if the software recognises the format. Images can be stored in numerous different formats. The three dominant formats for photographic digital images in most common usage are:

- ~ JPEG (.jpg) - Joint Photographic Experts Group
- ~ TIFF (.tif) - Tagged Image File Format
- ~ Photoshop (.psd) - Photoshop Document

JPEG - Industry standard for compressing continuous tone photographic images destined for the world wide web (www) or storage.

TIFF - Industry standard for images destined for publishing (magazines, newspapers and books etc.).

Photoshop - A default format used by the most popular image processing software.

Resolution

Digital images are constructed from square picture elements or '**pixels**'. Increasing the total number of pixels in an image at the scanning or capture stage increases both the quality of the image and its file size.
It is '**resolution**' that determines how large or small the pixels appear in the final printed image. The greater the resolution the smaller the pixels and the greater the apparent sharpness of the image. Resolution is stated in '**pixels per inch**' or '**ppi**'.

Note. With the USA. dominating digital photography, measurements in inches rather than centimetres are commonly used. 1 inch equals approximately 2.5 centimetres.

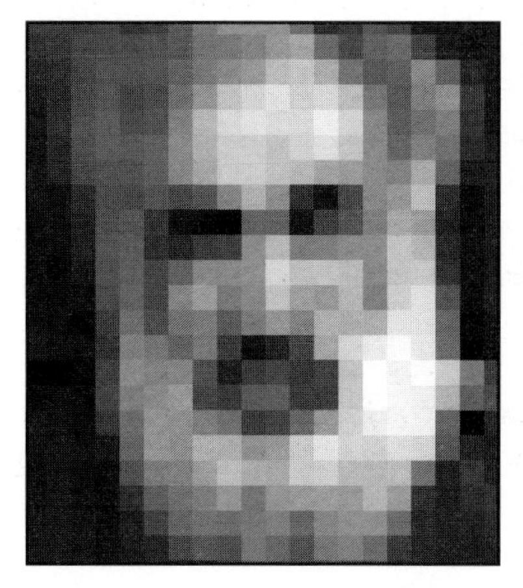

10 pixels per inch

20 pixels per inch

Example: If a 1 inch square postage stamp is scanned at 10ppi the digital image will have a total of 100 pixels (10 x 10 = 100). If the image is displayed '**same size**' as the original the output resolution will be stated as 10ppi and each pixel will measure 1/10 of an inch.
If the same stamp is scanned at 20ppi the image will now have 400 pixels. If displayed or printed 'same size' each pixel will now only be 1/20 of an inch.
Increasing the resolution will increase the quality of the image until it is limited by the output device. If the resolution exceeds that required by the printer to produce optimum quality the extra resolution is wasted.

Dots per inch

Printers place or transfer ink onto the paper in dots not squares. Higher quality printers put more dots of ink down per inch of paper. Printing devices therefore define the quality or '**output device resolution**' in '**dots per inch**' (dpi) not pixels per inch. The resolution of the image (measured in pixels per inch) is usually much smaller than the output device resolution used to print the image, i.e. each pixel is defined by more than one dot of ink.
Although there is a relationship between dpi and ppi they should not be confused during the print output stage, e.g. a 720dpi inkjet printer needs an image with a maximum resolution of only 240ppi (an RGB A4 digital file with a resolution of 720ppi would exceed 143Mb).

Resolution and print quality

When scanning or capturing an image it is essential to know how large and on what device you intend to print the image. This information is essential to ensure the maximum possible quality of your final image whilst keeping the file size as small as possible.
If the resolution is too high the handling of the file will be slow. If the resolution or file size is too small the quality of the final image will be low. If the resolution of the file is very low the individual pixels of the image will be visible and the image will be said to be '**pixellated**'.
If the intended print is very large, e.g. larger than A2, the resolution may be lowered without an apparent drop in sharpness. Large prints are usually viewed from further away and therefore require less resolution.

Changing print size from the same file

Resolution drops as the print size is increased (the pixels are spread further).
Example: A 4 x 5 inch print is scanned at 60ppi. If this digital image file is enlarged to 8 x 10 inches the resolution will drop to 30ppi. Both the original size image and the enlargement contain the same number of pixels but the quality of the larger image appears to be lower.

Note. A resolution of at least 150ppi is required to prevent pixellation.

104 pixels wide at 60ppi

104 pixels wide at 30ppi

Interpolation

If a digital image is captured with a fixed number of pixels the image is limited in both size and/or quality. It is possible with software to increase the number of pixels thus allowing an increase in size or resolution. Increasing the total number of pixels requires '**interpolation**'. New pixels are created from information obtained from the existing pixels. These new pixels are inserted between the existing pixels. These fabricated pixels cause a loss in overall quality of the image. If an image has to be scaled to fit a certain format or the specifications of a printer you should aim to scale down (reduce the number of pixels) rather than up.

Note. Avoid increasing the total number of pixels in a digital image whenever possible.

Scanning resolution

Scanning resolution is rarely the same as the resolution you require to print out your image. If you are going to create a print larger than the original you are scanning, the scanning resolution will be greater than the output resolution, e.g. a 35mm negative will have to be scanned in excess of 1200ppi if a 150ppi A4 colour print is required. If the print you require is smaller than the original the scanning resolution will be smaller than the output resolution.

The smaller the original the higher the scanning resolution.

In determining the file size and resolution required when scanning a negative, transparency or print you need to know the following:

~ Size and '**mode**' of the final image you require, e.g. A4 RGB, A3 Greyscale etc.
~ Optimum resolution of the image that can be utilised by the printer (proportional to, but not the same as, the dpi of the printer), e.g. 240ppi is usually the optimum resolution for a 720dpi inkjet printer.
Consult the printing company when the work will be commercially produced (see '**Calculating image resolution for output**').

Using the above information you can then do either of the following:

~ Create a 'dummy file' in your image processing software. Go to 'File' and select 'New...'. Type in the specifications (size, mode and resolution). This will give you a file size you can aim for when scanning. Simply increase the scanning resolution to obtain the appropriate file size.
~ Multiply the proportional change in size (original to required output size) by the resolution required by the printer, e.g. if an output size of 8 x 10 is required from a 4 x 5 original the dimensions have doubled (x2). If the printer requires a resolution of 150ppi the original should be scanned at 300dpi (2 x 150 = 300).

Note. Always scan with a slightly higher resolution or larger file size if you are unsure.

Size & Mode	Image Quality			
	72ppi low quality	150ppi medium quality	200ppi medium quality	300ppi high quality
A4 RGB	**1.43M**	**6.22M**	**11.10M**	**24.90M**
A4 Greyscale	**0.49M**	**2.07M**	**3.70M**	**8.30M**
A3 RGB	2.86M	12.44M	22.20M	49.80M
A3 Greyscale	1.98M	4.14M	7.40M	16.60M
A5 RGB	733K	3.11M	5.52M	4.14M
A5 Greyscale	245K	1.04M	1.84M	12.50M

File Size

Scanning

To maximise the quality of the final digital image it is worth spending some time getting the best possible scan from the print or film. This may require deselecting the 'auto' facility in the scanning software. The aim when scanning is to get the optimum tonal range possible before it is manipulated in the image processing software. Poor scans can be improved later but if the scan contains limited information the final image will be inferior to an image that was scanned properly in the first place. A good scan will have the following:

Sufficient pixels for the required image size and output device.

- The original's highlight and shadow detail preserved.
- Is colour corrected to that of the original.

To ensure a good scan an individual must have:

- Knowledge of the file size and/or scanning resolution (see '**Scanning resolution**').
- Clean original image and scanning device.
- Knowledge of the scanning software controls.

When the scanning software is opened the following should be checked before a '**preview**' is made:

- Correctly position the original within the scanning area of the device.
- Select the reflective (print) or transmissive (film) scanning mode.
- Select an appropriate image mode (RGB or Greyscale).
- Select '**descreen**' (image from magazine or book) or '**no descreen**' (photograph).

After the preview image appears the following needs to be carried out:

- Realign original if necessary (preview again if this is performed).
- Define scanning area using the rectangular marquee tool.
- Select scanning resolution and magnification until required file size is achieved.
- Adjust highlights, shadows and mid-tones using available controls.
- Adjust colour balance using available controls

Note. The inclusion of a 'grey card' (a photographic standard reflecting 18% of light) in the scanning area during a test scan may assist tonal and colour corrections.

Check the effectiveness of a scan by assessing the '**levels**' in the imaging software (see '**Digital manipulation - Levels - Altering the tonal range**)'. Highlights with detail and/or texture that will register in the final print should be between level 243 and 248 (anything brighter is likely to register as paper white). Shadows with detail and/or texture that will be seen in the final print should be between level 8 and 25 (anything darker is likely to register as solid black).
Neutral grey mid-tones in a colour image will register between 110 to 127 in all channels. Differences in the levels may indicate a colour cast.

Digital image output

If a digital image can be output as a hard copy, there are a number of different output devices including the following:

- Inkjet printers.
- Laser printers.
- Iris inkjet and dye sublimation printers.
- Image setters.

Inkjet and laser

Inkjet printers are by far the cheapest to purchase and most can now produce high quality prints (output device resolution of 720dpi or more). Inkjet printers are however slow and the cost of replacing inks is high (print for print) when compared to laser printers.
Quality of output is increased dramatically when using coated papers specifically designed for producing photographic images from inkjet printers.
Inkjet printers use cyan, magenta, yellow and black inks (CMYK) and the most sophisticated inkjet printers now have additional 'light cyan' and 'light magenta' inks for improved highlight quality. It is recommended that images sent to these printers should be left in RGB mode. Conversion to CMYK is best left to a skilled operative.

Iris inkjet and dye sublimation

These are high-end commercial quality printers and produce very high quality digital images. They are continuous tone printers and print with no visible dot structure and as a consequence the images produced can be used as original artwork for use in the printing industry.

Image setters

These are the commercial printers used to print newspapers, books and magazines. They utilise a '**half-tone**' dot structure very different to the dot structure utilised by inkjet printers. High quality books and magazines are printed using image setters with an output resolution of 2400dpi. Professional printers quote the quality in terms of a '**screen frequency**' which is stated in '**lines per inch**' or '**lpi**'. An image setter printing at 2400dpi may typically use a screen frequency of 175lpi. These image setters require colour images to be converted to CMYK.

Note. Images should only be converted from RGB to CMYK after consulting with the printing company being used.

Calculating image resolution for output

The image resolution required to create a hard copy is dependent on the output device being used, the size of the enlargement and the quality required by the individual.

A commonly used method for calculating the optimum image resolution that can be utilised by an inkjet or laser printer is to divide the output device resolution (ODR) by a factor of three, e.g. 720dpi divided by 3 = 240ppi. If the resulting file size is unmanageable by your computer it is possible to increase to a factor of four. The quality will still look reasonably good but will lower the working file size significantly.

If you are intending to send image files out for commercial printing consult the printer prior to creating the initial digital image file. It is very important to avoid having to increase the file size to accommodate a higher resolution required by the printer (see '**Interpolation**').

If an image setter is being used the printer may give you the screen frequency in lines per inch. If this is provided you should multiply this figure by a factor of 2, e.g. 133lpi multiplied by 2 = 266ppi.

Image resolution requirements of print output devices

Inkjet and laser: **Divide the output resolution by a factor of 3.**
Dye sublimation: **Image resolution is the same as quoted output resolution.**
Image setter: **Multiply the screen frequency by a factor of 2.**

Output file formats

When printing to inkjet and laser printers directly from the image processing software it is permissible to use any of the file formats supported by the software, e.g. Photoshop, TIFF, JPEG etc. If the file is to be imported by word processing or desktop publishing (DTP) software the file will usually have to be saved as either a TIFF or JPEG. If the desktop publishing file is to be printed commercially the standard file format is TIFF. Ensure that both the DTP file and all image files used in that file are kept together in one folder that can be easily accessed by the printer.

If an image is to be sent as an attachment to an email the file should be saved as a JPEG with the file extension '**.jpg**'. When saving the file in the image processing software the file should be saved without icons and previews to minimise the file size. If speed of transfer is required a high level of compression (low image quality) should be considered. Large file sizes may transfer very slowly via a modem.

If the image file is going to be used as part of a web page the JPEG format should again be used. A high level of compression is again required to minimise web page '**download time**'.

A file format that is being used increasingly for transferring DTP files to printers is the '**Portable Document Format**' or '**PDF**'. These can be created and viewed by the Adobe software '**Acrobat**'.

Digital

Manipulation

aims

- To offer an independent resource of technical information.
- To develop knowledge and understanding about the technical processes and procedures involved with manipulating digital image data.

objectives

- **Manipulate, modify and enhance** - a digital image file.

Introduction

After an image has been successfully captured and stored as a file format it can then be modified in preparation for printing. It may be necessary to do one of the following:

- ~ Change the mode, e.g. colour to black and white.
- ~ Crop the image to the required size.
- ~ Change the image resolution or sharpness so that it is suitable for the output device.
- ~ Alter the tonal range (contrast, brightness and levels).
- ~ Adjust colour hue and saturation.
- ~ Modify the content - remove, add or change the information.

Some of the above changes can be made at the scanning stage. The mode, colour balance and tonal range (levels of highlights and shadows) should be close to the desired outcome when initially scanned. Check information supplied with the scanning software for further details. To modify the content of the image after scanning or digital capture will require image processing software. Software is usually supplied with the scanning device but the industry standard '**Adobe Photoshop**' would normally be purchased separately. The software used will require a computer with sufficient performance to handle image processing. This is calculated in terms of '**memory**' and '**processor speed**'.

Computer memory

When an image file is opened by a computer it is held in the computer's short-term or working memory known as 'random access memory' or 'RAM'. The available RAM must exceed the image file size, i.e. a 6 megabyte image requires at least 6 megabytes of RAM to open the file. If the computer program that opens the file is used to change or modify this file additional RAM is required. A minimum of 32 RAM is required to produce colour prints up to A4 (64 RAM is recommended as a minimum for Adobe Photoshop). This is in addition to the RAM required to operate the '**system software**' of the computer ('Macintosh OS' or 'Windows'). RAM should not be confused with the computer's '**hard drive**' memory which is used to store the files and software once the computer is switched off.

Virtual memory and processor speed

It is possible to use some of the hard drive memory as RAM. This is called '**virtual memory**'. If virtual memory is used for image manipulation and enhancement the time required to process the data will greatly increase. Additional RAM can be purchased and installed in the computer.
The speed of the processor installed in the computer will also determine the time required to modify and manipulate image data. Pentium, PowerPC and G3 and G4 are some typical names of fast processors.

Image size

Before manipulation and enhancement can take place the '**image size**' must be adjusted for the intended output. This will ensure that optimum image quality and computer operating speed is achieved. Image size is described in three ways:

~ pixel dimensions (the number of pixels determines the file size in terms of kilobytes);
~ print size (output dimensions in inches or centimetres);
~ resolution (measured in pixels per inch or ppi).

If one is altered it will affect or impact on one or both of the others, e.g. increasing the print size must either lower the resolution or increase the pixel dimensions and file size. The image size is usually changed for the following reasons:

~ Resolution is changed to match the requirements of the print output device.
~ Print output dimensions are changed to match display requirements.

When changing an image's size a decision can be made to retain the proportions of the image and the pixel dimensions. These are controlled by the following:

~ If '**Constrain Proportions**' is selected the proportional dimensions between image width and image height are linked. If either one is altered the other is adjusted automatically and proportionally. If this is not selected the width or height can be adjusted independently of the other and may lead to a distorted image.

~ If '**Resample Image**' is selected adjusting the dimensions or resolution of the image will allow the file size to be increased or decreased to accommodate the changes. Pixels are either removed or added to accommodate the change in resolution or print size. If deselected the print size and resolution are linked. Changing width, height or resolution will change the other two.

Change resolution and retain print size

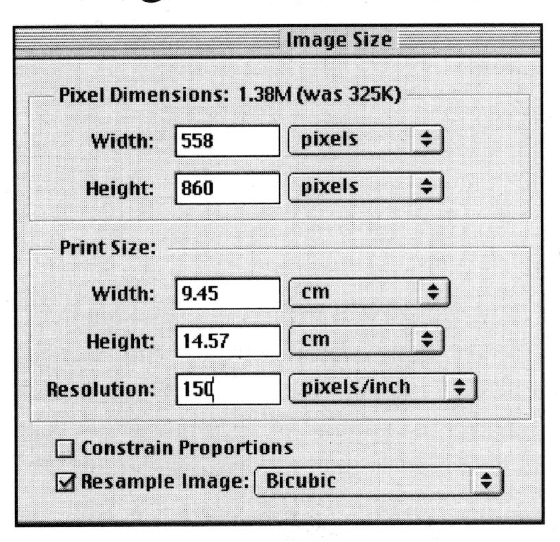

Aim:
To increase or decrease resolution whilst keeping the print size the same.
Result:
Pixel dimensions change leading to a reduced or enlarged file size.
Action:
Check 'Resample Image' box. If left unchecked the print size would change in order to keep the file size constant. Type in the revised resolution. Interpolation is required to increase the file size.

Change print size and retain resolution

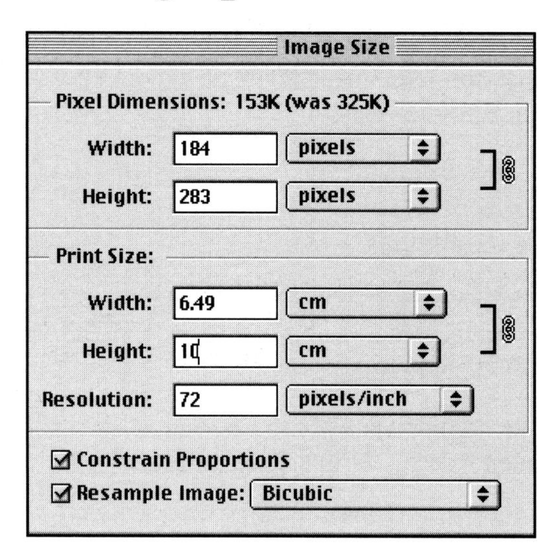

Aim:
To change the print size whilst retaining the original resolution.
Result:
Pixel dimensions alter leading to a reduced or enlarged file size.
Action:
Check 'Constrain Proportions' to reduce or increase height and width proportionally). Check 'Resample Image' to prevent resolution rising. Type in the revised height or width.

Change print size and maintain file size

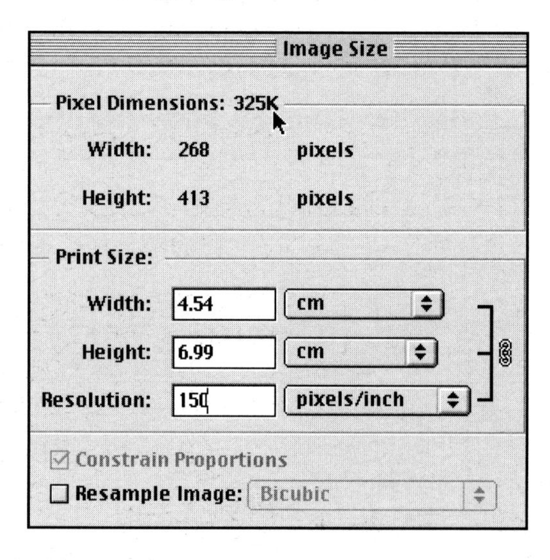

Aim:
Modify output size or resolution whilst maintaining the file size.
Result:
Resolution and print size are linked. Changing one will affect the other.
Action:
Uncheck the resample box . This now links resolution and print size ('Constrain Proportions' is automatically selected). The file size remains the same whatever the change.

Resampling images

Resampling an image so that the file size increases will lower the visual quality. If this is necessary use the '**Bicubic**' option in the resample preferences and limit the increase to double the original size to minimise the loss in quality. Always use the '**Unsharp Mask**' after resampling rather than before and restrict the amount of resampling that is performed on a single image. If the software allows the operator to crop, resize and rotate the image at the same time this function should be utilised whenever possible.

Altering the tonal range

Using image processing software the tonal range of a digital image (or a localised area within the image, see '**Selection tools and techniques**') can be controlled in a number of ways. These include:

- Brightness and contrast
- Levels
- Curves

Brightness and contrast is the basic control and is easy to understand for anyone already familiar with exposure and paper grades in black and white photography. '**Adjust Levels**' offers more control, whilst '**Curves**' offers the most sophisticated control over the tonal range of a digital image (also most difficult to master). It is recommended that once an image has been cropped and resized the tonal range is adjusted using the 'Adjust Levels'.

Levels

Each channel in a digital image is divided into 256 '**levels**' of brightness (0 to 255). The darkest level is 0 and the brightest is 255. A good scan of an average image will record pixels in all 256 levels. Many scans, however, will lack information in highlights or shadows. If '**Auto Levels**' is selected the available pixels are distributed over the full range from 0 to 255. Although this is a very quick way of adjusting the tonal range it offers extremely limited control over the precise look of the image and is not suitable for many images.
If 'Adjust Levels' is selected a '**histogram**' or graph appears. The vertical lines represent the number of pixels for each level. If an image has a broad spread of levels containing pixels it is said to have a broad dynamic range. Image should ideally have a broad dynamic range if they are to be manipulated or enhanced without suffering excessive loss of quality.

Bicycle - Jana Liebenstein

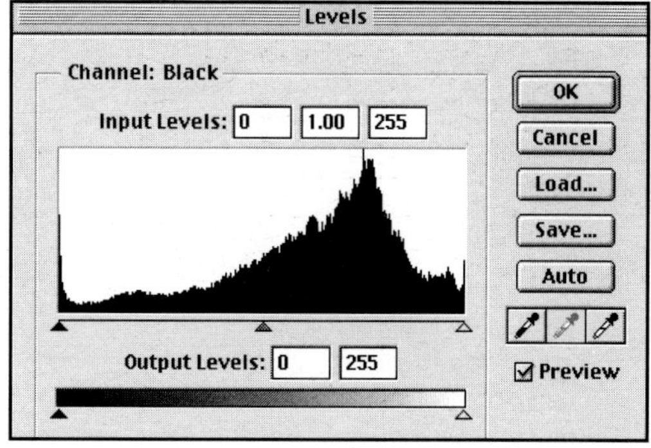

Histogram of Bicycle

The histogram of the bicycle above shows a dynamic range of tones with dominant light tones gathered towards the higher levels. Directly beneath the histogram are three '**sliders**' (black point, mid-point and white point). Control over the tonal range can be exercised by clicking and dragging these sliders to control shadows, mid-tones and highlights.

Adjusting the levels

The adjustment of levels using a computer software program can greatly enhance the visual quality of an image. The finest quality digital images are, however, produced from digital files (sourced from a camera or scanning device) that have a broad dynamic range. This dynamic range is dependent on correct exposure in the camera and/or careful scanning. The levels for the highlights with detail should be between 200 and 250. The levels for the shadows with visible detail should be above 5 (this can be checked using the information palette). For this reason not all images should have both a white and black point (see image below).

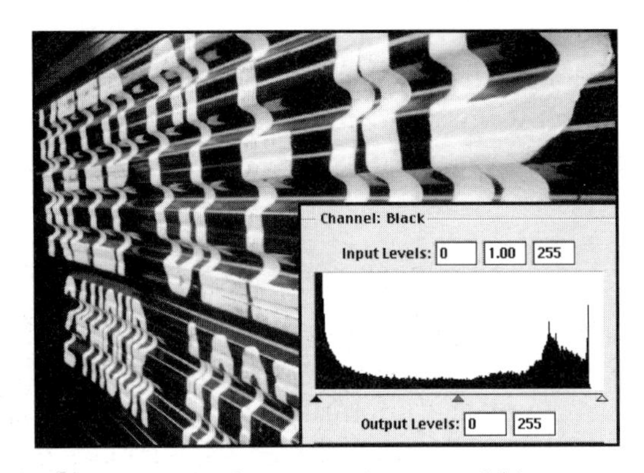

High contrast image and histogram

A high contrast image with low pixel totals in the mid-tones. The highest levels show a complete absence of pixels. The white point slider could be moved to the left to further increase the contrast but detail would be reduced in the highlights.

Low contrast image and histogram

This low contrast image is due to a poor scan and the resulting dynamic range is narrow. The dynamic range can be improved using the sliders but an improved scan should be the preferred option.

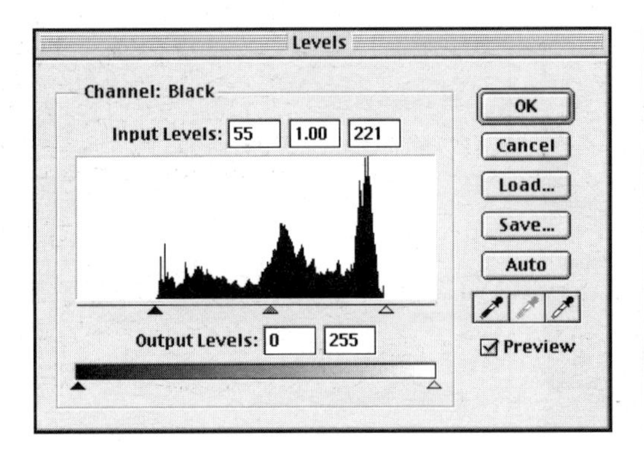

Improving the tonal range using the sliders

To improve a narrow dynamic range click and drag the black and white point sliders to the first levels showing pixels. Moving the centre mid-point slider will lighten or darken the mid-tones.

The end result

The results of the tonal adjustments show an image with a broad range of tones and increased contrast. The software cannot, however, replace missing information in the highlights and shadows.

Modifying content

The most commonly used tools in Photoshop are the retouching tools and selection tools. Once an image is cropped, resized and the tonal range has been corrected the individual may want to modify the content. This may be a simple act of retouching, removing dust and scratches or may entail adding or removing subject matter. Just as in word processing, it is possible to cut, copy and paste information from one area of the document to another. These are carried out using the following tools:

~ Rubber stamp
~ Marquee
~ Lasso
~ Magic wand

Retouching 

The primary tool for localised retouching is the '**rubber stamp**' tool. It is able to paint with pixels selected from another part of the image. These pixels can be tones, colours or textures required to disguise blemishes or replace subject matter. The tool is quickly able to remove hairs and dust by simply selecting a tone adjacent to the blemish and then dragging the tool icon over the area to be changed.

Master image

Manipulated image

The tool is often referred to as a '**cloning tool**' because rather than removing pixels it is cloning groups of pixels from another area. With care it is possible to duplicate an entire subject within the image. The image above demonstrates how a landscape composition has been manipulated to fit a portrait format. This has been achieved by cloning the sign and the life-ring and moving them to the right. The original life-ring is then removed.

Creating composite images

One of the most skilled areas of digital manipulation is the ability to make accurate selections of pixels for repositioning, modifying or exporting to another image. This skill allows convincing manipulations and composite images to be created. Obvious distortions of photographic originals are common in the media but so are images that are subtle and undetectable. These images are used every day but are not credited as having been changed radically from the original.
Selections are made for a number of reasons:

- Making an adjustment or modification to a localised area, e.g. colour, contrast etc.
- Defining a subject within the overall image to move or replicate.
- Defining an area where an image or group of pixels will be inserted ('paste into...').

Selection tools and techniques

A number of selection tools are available for selecting groups of pixels in very different ways. Pixels can be selected by placing a defined shape around them using the rectangular and elliptical '**marquee tools**', drawn around by a '**lasso tool**' and selected by their colour or tone using the '**magic wand tool**'.
When cutting or copying squares, oblongs, circles or ovals the rectangular or elliptical marquee tools are used to define the selection by clicking the mouse and dragging a shape (indicated by a moving dotted line or '**marching ants**') over the area required.
The lasso is used to draw around a section of the image whilst the magic wand is placed on a specific area of the image and the mouse clicked. The magic wand selects adjacent pixels that are either the same or similar to the one specified by the click of the mouse.
The selection tools can be used individually or combined. A selection may be made and then added to or subtracted from by depressing keyboard keys when the selections are made. Using Photoshop the'Shift' key should be depressed to add a subsequent selection to the existing selection and the 'Option' key depressed whilst subtracting a selection from the existing selection.
Selections can then be adjusted independently from the rest of the image, copied (leaving the original image intact) or cut, using either keyboard short-cuts or the commands from the 'Edit' menu.

Layers

In Adobe Photoshop (version 3 onwards) new selections that have been either copied, or cut, can be pasted onto new '**layers**'. Layers can be likened to clear acetate sheets, each with part of an image, placed one above another to create a composite image. The layers can be rearranged and the opacity of each layer changed so that different visual outcomes can be achieved.

Selection options

The ability to create a composite image that looks subtle, realistic or believable rests in the ability of the viewer to detect where one image starts and the other finishes. The edges of each selection can be modified so that it appears as if it belongs, or is related, to the surrounding pixels.
Options are available with most image processing software to alter the appearance of the edges of a selection. Edges can appear sharp or soft (a gradual transition between the selection and the background). The options to effect these changes are:

- Feather
- Anti-aliasing
- Defringe

Feather

When this option is chosen the pixels at the edges of the selection are blurred. The edges are softer and less harsh. This blurring may either create a more realistic montage or cause loss of detail at the edge of the selection.
You can choose feathering for the marquee or lasso tools as you use them, or you can add feathering to an existing selection. The feathering effect only becomes apparent when you move or paste the selection to a new area.

Anti-aliasing

When this option is chosen the jagged edges of a selection are softened. A more gradual transition between the edge pixels and the background pixels is created. Only the edge pixels are changed so no detail is lost. Anti-aliasing must be chosen before the selection is made. It cannot be added afterwards.

Defringe

When a selection has been made using the anti-alias option some of the pixels surrounding the selection are included. If these surrounding pixels are darker, lighter or a different colour to the selection a fringe or halo may be seen.
A defringe command, if available, replaces the different fringe pixels with pixels of a similar hue, saturation or brightness found within the selection area.

Saving selections

It is possible in some image processing software to save selections and images with multiple layers. A selection can be saved as a channel and recalled at a later date when the selection is required for use. An image with layers may only be saved as a Photoshop file (.psd) when using the image processing software 'Photoshop'. An image with saved selections cannot be saved as a JPEG.

The musician is selected using a combination of the magic wand and lasso tools. The selection is contracted by two pixels and a one pixel feather is then applied (to soften edge).

The selection is cut (or copied) and then pasted onto a new image. It is important to ensure that both images are approximately the same mode size and resolution.

The new background was created by changing the focal length of a zoom lens during a 1/8 second exposure.

The final image has more impact with a greater sense of movement than the original which appears static in comparison.

Blending images

Blending two images in the computer is similar to creating a double exposure in the camera or sandwiching negatives in the darkroom. Using image processing software the individual can exercise a greater degree of control over the final outcome. This is achieved by controlling not only the position and opacity of each layer but also which areas of the image will be blended and which will remain untouched.

It is a technique whereby the texture or pattern from one image can seem to model the form of another.

Blend

The following operations were performed in Adobe Photoshop to achieve the blended image above. This is a suggestion only to indicate the type of effects that can be achieved using a variety of image processing software.

1. Select or create one image where a three-dimensional subject (preferably with a smooth surface texture) is modelled by light. Scan the image and save as 'form'.
2. Select or create another image where the subject has an interesting and bold texture or pattern. Scan the image and save as 'texture'.
3. In '**image size**' check that the image size and resolution for each image are similar.
4. Open image labelled 'form' and make a selection of the subject using the magic wand, marquee and lasso tools. Save this selection.
5. Open the image labelled 'texture', choose '**select all**' (from the select menu) and '**copy**' (from edit menu).
6. Make active (click on the window) the image 'form' and choose '**paste into...**' (from the 'edit' menu). This action creates a new layer and a '**layer mask**'.
7. Select this new layer in the layers palette and the '**multiply**' option from the layers palette. Adjust (decrease) the opacity of this layer until the required effect is obtained.

Increasing image sharpness

After a digital image had been enhanced, modified or manipulated it may be necessary to increase the apparent sharpness of the image as the last step prior to output. Many images benefit from some sharpening prior to output even if they were photographed and scanned with sharp focus. The filter used by Adobe Photoshop that controls image sharpness is the '**Unsharp Mask**' which can be found in the '**filters**' menu. The unsharp mask increases apparent sharpness by emphasising the edges between different tones within the image (chemical developers are used with film to create similar effects). If a light tone is next to a dark tone, the edge between the two can be emphasised digitally by lightening the pixels along the edge of the light side and darkening the pixels along the edge of the dark side. It appears that dark and light lines are drawn between areas of tone.
The three controls over sharpness are amount, radius and threshold.

Amount - controls the increase in **contrast** between different tones at the edges (how dark or light the edges become). 80 to 180% is normal. The amount is usually lower when sharpening images of people.

Radius - controls the **width** of the effect occurring at the edges. There is usually no need to exceed 1 pixel.

Threshold - controls **where** the effect takes place. A low threshold affects all tonal differences whereas a high threshold affects only edges with a high tonal difference. The threshold is usually kept very low (0 to 2) for images captured with digital cameras or scanned from medium and large format film. The threshold is usually set at 3 with images from 35mm. Threshold is increased to avoid accentuating grain, especially in skin tones.

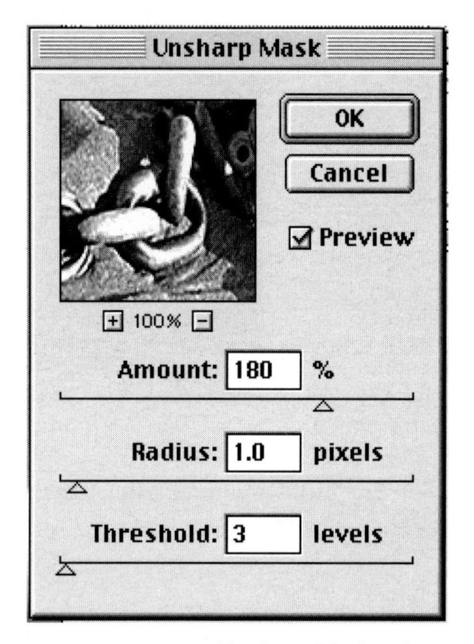

Unsharp Mask palette

Before the unsharp mask is applied

After unsharp mask is applied. Radius has been set at 3 pixels to exaggerate edge emphasis

Resources

Michael Wearne

aims

- To offer resources for the implementation of a photographic curriculum.
- To provide work sheets to help students organise their planning.
- To provide progress reports to enable teachers to document learning and monitor student progress.

contents

- Darkroom design
- A sample controlled test
- Student progress report sheet
- Student work sheet
- A sample summer project

Darkroom design

Constructing a darkroom does not have to be expensive or complicated. An available room can be modified quickly and effectively if the following are present:

1. A sink with running water.
2. An electricity supply.
3. An extractor fan to supply adequate ventilation.

Preparing the room

The dry surface - A dry surface is needed where the enlargers have access to a power supply. In an ideal environment each enlarger has its own power supply just above the work surface. Each enlarger needs about one metre of available work surface and these can be partitioned off up to the height of the enlarger being used. Although this is useful to prevent the light from one student's enlarger fogging the paper of another, it is by no means essential if students take care to switch off their enlarger whilst removing the negative carrier.

The wet surface - A wet surface for the processing of printing paper is also required. If the money is available a flat bottomed PVC sink long enough to contain the four processing trays on a rack placed inside the sink is ideal but not essential. The first three trays can be placed directly onto a washable surface and the final wash tray can be placed in the adjacent sink where it can have access to running water via a flexible hose. The water is allowed to overflow this final tray in the washing process. Precautions can be taken to avoid flooding caused by small pieces of photographic paper or paper towels blocking the plug hole by inserting a small plastic tube into the plug hole thereby raising the water outlet to the sink.

The entrance - The most convenient entrance to a darkroom is a light trap painted matt black. This allows people to enter and leave the darkroom without the need for others to put away light sensitive materials. The alternatives to a light trap are double doors or a revolving door which although less convenient will take up less space.

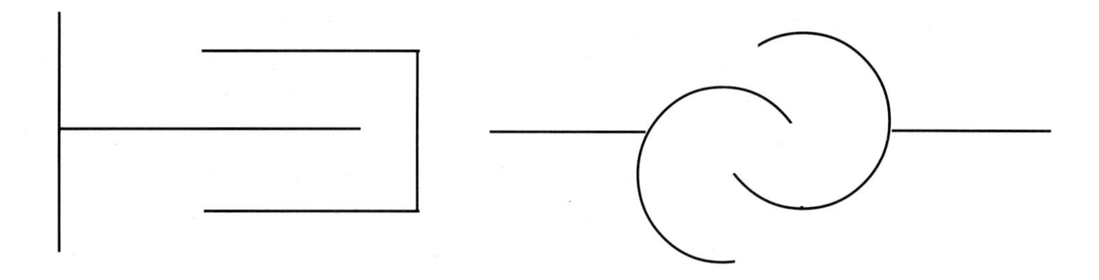

Two examples of a darkroom light trap

Ventilation - A light trap will allow free circulation of air but doors must be provided with a light tight grill. The extractor fan must also be light proof. Any materials used to prevent direct light entering the darkroom should be painted with matt black paint.

Safelight - The most convenient form of safelight available are special red fluorescent tubes that fit directly into the existing fittings. If more than one fitting is available on separate switches one can be left with the original tube to provide bright illumination for preparation of chemicals and clearing up at either end of the day. Safelights can also be bought which screw into existing bayonet type fittings. When choosing the type of safelight you will also need to consider whether you will be handling certain types of film which can also be handled in safelight conditions. If you are not sure as to which safelight you will need for the long term you are advised to choose red lights rather than orange as these will allow students to use all black and white materials that can be handled in safelight conditions.
Towel dispenser - Students will need the use of towels to dry hands after washing and mop any spillage of water or chemicals.

Essential equipment

Enlargers - Enlargers should be bought that can use filters to alter the contrast of multigrade paper. These can either be equipped with a multigrade head, colour head or have a filter draw. When purchasing the enlarger enquiries should be made as to whether it comes with a 50mm lens suitable for enlarging from a 35mm negative.
Developing trays, printing tongs and rubber gloves - The trays should be slightly larger than the printing paper you intend to use. The printing tongs and rubber gloves are essential to allow students to transfer prints which are wet with chemicals between trays.
Mixing cylinders and funnel - These are necessary for diluting the concentrate and returning diluted chemicals to containers for storage.
Eye protection - Students must wear eye protection if they are expected to pour chemicals.
Clock - A large wall mounted clock with a sweeping second hand should be mounted on a wall that is easily visible by all the students working in the darkroom. This will be of use for timing operations where there is a lack of individual timers and will also make sure that students do not overrun photographic sessions.

Non-essential equipment

Enlarger timers - These are useful and accurate but again are not essential. In-line switches on the power cables to the enlargers can be fitted to switch the enlarger light on and off. Students can be taught to use exposure times longer than 10 seconds and count accurately by speaking the same word between each number, e.g. one-potato-two-potato-three etc.
Printing easels - These are used to hold printing paper flat and in the right position on the baseboard. They are, however, a luxury and are easily damaged. Modern resin coated printing paper will lie flat without the use of an easel. A separate board with a white sticky-back plastic oblong the same size as the printing paper can be used to frame the projected image and provide positioning for the photographic paper. If borders are required on the photographic paper a window mount can be cut from card and used together with a baseboard which will register both the paper and the window mount.
Supplementary equipment - Focus finders, rotary trimmer, drying rack. These make life easier in the darkroom but can be added after the darkroom is already established.

Controlled test

Your completed work must include contact sheets and research work as well as your finished presentation piece. You should give clear evidence of how you have approached your work and how your ideas have developed during the course of the test. Any experimentation into style and technique should be recorded clearly on the research sheets. The final work should be mounted on A1 or A2 card.

Note. You are encouraged to make full use of the range of techniques and skills covered by the course wherever appropriate.

Answer ONE question only.

1. **Manual labour** You have been commissioned by the council to present a visual documentary about manual labour. Your piece could focus on a man, woman, group or team. Your work should emphasise the interaction between people and their surroundings, with particular reference to the supporting objects and images associated with their occupations and pastimes.

2. **Theatre workshop** Produce a set of promotional photographs for a theatre workshop that shows the company are encouraged to develop a wide variety of acting styles. Your photographs should demonstrate a variety of moods and emotions using one or more people. It is recommended that you stage the lighting and poses specifically for the camera. The moods and emotions might include happiness, sadness, curiosity, contemplation, boredom, excitement, friendliness, hostility, arrogance, delight, fear, satisfaction, anticipation, anger, peace, concentration, uncertainty, and frustration.

3. **Pressure group** Produce a set of promotional photographs that would be suitable for use by a pressure group, e.g. Greenpeace, Friends of the Earth, Campaign for Nuclear Disarmament, Anti Nazi League or one of your own choice. The images you produce should promote the principal objectives or symbolism related to the particular group.

4. **Illustration** Produce a set of images that illustrates a poem, book extract, song lyric or newspaper article. You may incorporate the text into the finished piece of work.

5. **Surreal story** Produce a sequence of images that communicates a surreal short story. The source of your imagery could be a recent dream or daydream that you may have experienced.

6. **Hands** Photograph hands in expressive postures or engaged in interesting activities. You may photograph one hand by itself, both hands of one person or the hands of several people together. Do not include a full face with the hands, although part of a face is acceptable.

7. **Still life** Produce a set of still life photographs that explores a range of old objects, things that are worn out from age or use. The final work can be either abstract or commercial.

8. **Shadows** Photograph an object (or part of it) along with its shadow, exploring how the shadow can add visual interest to an image.

Progress report

Student Name .... Class .... Date .... Time..........

Assignment feedback

Creativity ....
Composition ....
Presentation ....

Exposure ....
Processing ....
Focusing ....
Contrast ....

Attendance .... Punctuality .... Deadlines ....

General progress

Creative approach to design activities ....
Management of time ....
Technical competency ....
Appropriate use of the study guides to aid research ....
Points raised by student

....

....

Points raised by tutor

....

....

Negotiated statement on progress

....

....

....

Necessary action agreed to be undertaken

....

....

....

Signature of student ....

Signature of tutor .... **Date** ....

Work sheet

Student Name ....Year .... Group ....
Assignment ....

Research

1. What is it you like about the images you have been looking at? ....
....
....
2. What techniques has the photographer used? ....
....
....

Preliminary work

1. Which images do you like from your first proof sheets? ....
....
....
2. How could you improve upon these images? ....
....
....

Plans and ideas

1. What elements from your research can you use in your own work?...............
....
....
....

2. What ideas do you have to develop a theme? ....
....
....
....
....

Organisation of final shoot

Location .... Time and date ....
Estimated time needed for shoot ....
Permission or tickets required .... Contact name ....
Equipment needed ....
Props needed ....

Summer project

Introduction

The word 'summer' can conjure up many different images: strawberries and cream, hot lazy days, rain soaked day trippers, excited holidaymakers, bored school children, cricket on the village green, discarded ice-creams in the sand, littered beaches etc.

Assignment

In this assignment you must record one aspect of your summer this year, something that has attracted your visual curiosity. Do not necessarily choose a stereotypical view of summer, i.e. a picture postcard shot of the beach or family snap shots.
The images that you choose should relate with each other in some way and portray a theme or aspect of summer that you would like to convey. The photographs may be abstract, documentary shots from real life or shots that have been set up to convey a message.
You may work in black and white or colour and will be expected to have taken at least two rolls of film. It is not expected that you will process or print your own images over the summer. You may have your films processed and printed commercial. If you decide to work in black and white you will find that the film Ilford XP2 or Kodak T 400 CN can be processed quickly and cheaply through most photographic high street minilabs.
Edit your prints down to six images and present them on one sheet of card no larger than A3 in size. The deadline for this project is the first week of term.

Assessment criteria

Your work will be assessed using the following criteria:

1. Arrangement and composition of subject matter within the frame.
2. Use of light to create appropriate mood.
3. Sympathetic use of viewpoint and creative use of shutter speed.
4. Visual clarity of idea and theme.
5. Presentation.

Gallery

Shaun Guest

Shaun Guest

Shaun Guest

Natalie Wright

Natalie Wright

Shane Bell

Georgia Tipperman

Glossary

Aliasing The display of a digital image where a curved line appears jagged due to the square pixels.

Ambient light The natural or artificial continuous light that exists before the additional lighting is introduced.

Analyse/Analysis To examine in detail.

Anti-aliasing The process of smoothing the appearance of a curved line in a digital image.

Aperture A circular opening in the lens that controls light reaching the film.

Backlit A subject illuminated from behind.

Balance A harmonious or stable relationship between elements within the frame.

Bit Short for binary digit, the basic unit of the binary language.

Blurred An image or sections of an image that are not sharp. This can be caused through inaccurate focusing, shallow depth of field or a slow shutter speed.

Bracketing Over- and underexposure either side of a meter-indicated exposure.

B setting The shutter speed setting B allows the shutter to stay open as long as the shutter release remains pressed.

Bounced light Lighting that is reflected off a surface before reaching the subject.

Byte 8 bits. The standard unit of binary data storage containing a value between 0 and 255.

Cable release A cable that allows the shutter to be released without shaking the camera when using slow shutter speeds.

Camera shake Blurred image caused by camera movement during the exposure.

CCD Charge coupled device. A solid state image pick-up device used in digital image capture.

Channels A method of separating a digital colour image into primary or secondary colours.

Cloning tool A tool used for replicating pixels in digital photography.

Close-up lens A one element lens that is attached to the front of the camera's lens. This allows the image to be focused when the camera is close to a subject.

Close down A term referring to the action of making the lens aperture smaller.

CMYK Cyan, magenta, yellow and black. The inks used in four colour printing.

Composition	The arrangement of shape, tone, line and colour within the boundaries of the image area.
Compression	A method of reducing the file size when a digital image is closed.
Constrain proportions	Retain the proportional dimensions of an image when changing the image size.
Contact print	A print created by placing objects or negatives in direct contact with photographic paper and exposing this paper to light.
Context	The circumstances relevant to something under consideration.
Contrast	The difference in brightness between the darkest and lightest areas of the image or subject.
CPU	Central processing unit used to compute exposure.
Crop	Reduce image size to enhance composition or limit information.
Curves	Control for adjusting tonality and colour in digital photography.
Decisive moment	The moment when the arrangement of the moving subject matter in the viewfinder of the camera is composed to the photographer's satisfaction.
Dedicated flash	A flash unit that is fully linked to the camera's electronics and uses the camera's own TTL light meter to calculate correct exposure.
Defringe	The action of removing the edge pixels of a selection.
Density	The measure of opacity of tone on a negative.
Depth of field	The zone of sharpness variable by aperture, focal length or subject distance.
Diagonal	A slanting line that is neither horizontal nor vertical.
Differential focusing	Use of focus to highlight specific subject areas
Diffused light	Light that is dispersed (spreads out) and is not focused.
Diffuser	Material used to disperse light.
Digital image	A computer-generated photograph composed of pixels (picture elements) rather than film grain.
Diminishing perspective	A sense of depth in a two-dimensional image provided by the reduced size of subjects as they recede into the distance.
Dioptres	Term for close-up lenses.
Dissect	To cut into pieces. The edge of the frame can dissect a familiar subject into an unfamiliar section.
Dpi	Dots per inch. A measurement of resolution.
Dynamic tension	An image which lacks either balance or harmony and where visual elements cause the eye to move out of the image.
DX coding	Bar coded film speed rating.
Edit	Select images from a larger collection to form a sequence or theme.
Emulsion	Light sensitive layer on film or paper.

Evaluate Assess the value or quality of a piece of work.

Exposure Combined effect of intensity and duration of light on a light sensitive material.

Exposure compensation To increase or decrease the exposure from a meter-indicated exposure to obtain an appropriate exposure.

Exposure meter Device for the measurement of light.

Extreme contrast A subject brightness range that exceeds the film's ability to record detail in all tones.

F-numbers A sequence of numbers given to the relative sizes of aperture opening. F-numbers are standard on all lenses. The largest number corresponds to the smallest aperture and vice versa.

Fast film A film that is very light sensitive compared to other films. Fast film has a high ISO number and can be used when the level of ambient light is low.

Feather The action of softening the edge of a digital selection.

Field of view The area visible through the camera's viewing system.

Figure and ground The relationship between subject and background.

Fill flash Flash used at a reduced output to lower the subject brightness range.

Fill Use of light to increase detail in shadow area.

Film speed A precise number or ISO rating given to a film indicating its degree of light sensitivity. See 'Fast film' and 'Slow film'.

Filter Treated or coloured glass or plastic placed in front of the camera lens to alter the transmission of different wavelengths of light thereby altering the final appearance of the image.

Filter factor A number used to indicate the effect of the filter's density on exposure.

Flare Unwanted light, scattered or reflected within the lens assembly creating patches of light and degrading image contrast.

Focal length Distance from the optical centre of the lens to the film plane when the lens is focused on infinity. A long focal length lens (telephoto) will increase the image size of the subject being photographed. A short focal length lens (wide-angle) will decrease the image size of the subject.

Focal plane shutter A shutter directly in front of the film plane.

Focal point Point of focus at the film plane or point of interest in the image.

Focusing The action of creating a sharp image by adjusting either the distance of the lens from the film or altering the position of lens elements.

Format The size of the camera or the orientation/shape of the image.

Frame The act of composing an image. See 'Composition'.

Golden section A classical method of composing subject matter within the frame.

Grain	Tiny particles of silver metal or dye which make up the final image. Fast films give larger grain than slow films. Focus finders are used to magnify the projected image so that the grain can be seen and an accurate focus obtained.
Grey card	Card which reflects 18% of incident light. The resulting tone is used by light meters as a standardised mid-tone.
Guide number	A measurement of the power of flash relative to the film speed being used. Divide the guide number by the maximum aperture available to find the maximum working distance in metres for 100 ISO film.
Half-tone	A system of reproducing the continuous tone of a photographic print by a pattern of dots printed by offset litho.
Hard copy	A print.
Hard drive	Memory facility which is capable of retaining information after the computer is switched off.
Hard light	A light source which appears small to the human eye and produces directional light giving well-defined shadows, e.g. direct sunlight or a naked light bulb.
High key	An image where light tones dominate.
Highlight	Area of subject receiving highest exposure value.
Histogram	A graphical representation of a digital image indicating the pixels allocated to each level.
Horizontal	A line that is parallel to the horizon.
Hot shoe	Plug in socket for on-camera flash.
Incident light reading	A measurement of the intensity of light falling on a subject.
Interpolation	A method of increasing the apparent resolution of an image by adding pixels of an average value to adjacent pixels within the image.
ISO	International Standards Organisation. A numerical system for rating the speed or relative light sensitivity of a film.
Infrared film	A film which is sensitive to wavelengths of light longer than 720nm but which are invisible to the human eye.
JPEG (.jpg)	Joint Photographic Experts Group. Popular image compression file format.
Juxtapose	Placing objects or subjects within a frame to allow comparison.
Key light	The main light casting the most prominent shadows.
Kilobyte	1024 bytes.
Lasso tool	Selection tool used in digital editing.
Latitude	Ability of the film to record the brightness range of the subject.

Layers	A composite digital image where each element is on a separate layer or level.
LCD	Liquid crystal display.
LED	Light-emitting diode. Used in the viewfinder to inform the photographer of exposure settings.
Lens	An optical device usually made from glass that focuses light rays to form an image on a surface.
Levels	The method of assigning a shade of lightness or brightness to a pixel.
Light meter	A device that measures the intensity of light so that the optimum exposure for the film can be obtained.
Long lens	Lens with a reduced field of view to normal.
Low key	An image where dark tones dominate.
Macro	Extreme close-up.
Magic wand tool	Selection tool used in digital editing.
Marching ants	A moving broken line indicating a digital selection of pixels.
Matrix metering	A reflected light meter reading which averages the readings from a pattern of segments over the subject area. Bias may be given to differing segments according to preprogrammed information.
Marquee tool	Selection tool used in digital editing.
Maximum aperture	Largest lens opening.
Megabyte	A unit of measurement for digital files. 1024 kilobytes.
Megapixels	More than a million pixels.
MIE	Meter-indicated exposure.
Minimum aperture	Smallest lens opening.
Mode (digital image)	RGB, CMYK etc. The mode describes the tonal and colour range of the captured or scanned image.
Multiple exposure	Several exposures made onto the same frame of film or piece of paper.
Negative	An image on film or paper where the tones are reversed, e.g. dark tones are recorded as light tones and vice versa.
Neutral density filter	Reduces the amount of light reaching the film, enabling the photographer to choose large apertures in bright conditions or extend exposure times.
Objective	A factual and non-subjective analysis of information.
ODR	Output device resolution.
Opaque	Not transmitting light.
Open up	Increasing the lens aperture to let more light reach the film.
Overall focus	An image where everything appears sharp.
Pan	To follow a moving subject.

Perspective	The apparent relationship of distance between visible objects thereby creating the illusion of depth in a two-dimensional image.
Perspective compression	Flattened perspective created by the use of a telephoto lens and distant viewpoint.
Photoflood	Tungsten studio lamp with a colour temperature of approximately 3400K.
Pixel	The smallest square picture element in a digital image.
Polarising filter	A grey looking filter used to block polarised light. It can remove or reduce unwanted reflections from some surfaces and can increase the colour saturation and darken blue skies.
Portrait lens	A telephoto lens used to capture non-distorted head and shoulder portraits with shallow depth of field.
Portrait mode	A programmed exposure mode that ensures shallow depth of field.
Previsualise	The ability to decide what the photographic image will look like before exposure.
Processor speed	The capability of the computer's CPU measured in megahertz.
Pushing film	The film speed on the camera's dial is increased to a higher number for the entire film. This enables the film to be used in low light conditions. The film must be developed for a longer time to compensate for the underexposure.
Push-processing	Increasing development to increase contrast or to compensate for underexposure of films that have been rated at a higher speed than recommended.
RAM	Random access memory, the computer's short-term or working memory.
Reflector	A surface used to reflect light in order to soften harsh shadows.
Refraction	The change in direction of light as it passes through a transparent surface at an angle.
Resample image	Alter the total number of pixels describing a digital image.
Resolution	A measure of the degree of definition, also called sharpness.
RGB	Red, green and blue. The three primary colours used to display images on a colour monitor.
Rubber stamp	A tool used for replicating pixels in digital imaging.
Rule of thirds	An imaginary grid that divides the frame into three equal sections vertically and horizontally. The lines and intersections of this grid are used to design an orderly composition.
Saturation (colour)	Intensity or richness of colour hue.
Scale	A ratio of size.

Selective focus	The technique of isolating a particular subject from others by using a shallow depth of field.
Self-timer	A device which delays the action of the shutter release. This can be used for extended exposures when a cable release is unavailable.
Sharp	In focus. Not blurred.
Shutter	A mechanism that controls the accurate duration of the exposure.
Shutter-priority	Semi-automatic exposure mode. The photographer selects the shutter speed and the camera sets the aperture.
Silhouette	The outline of a subject seen against a bright background.
Skylight filter	Used to reduce or eliminate the blue haze seen in landscapes. It does not affect overall exposure so it is often used to protect the front lens element from damage.
Sliders	A sliding control in digital editing software used to adjust colour, tone, opacity etc.
Slow film	A film that is not very sensitive to light when compared to other films with a higher ISO number. The advantage of using a slow film is its smaller grain size.
SLR camera	Single Lens Reflex camera. The image in the viewfinder is essentially the same image that the film will see. This image, prior to taking the shot, is viewed via a mirror which moves out of the way when the shutter release is pressed.
Soft light	This is another way of describing diffused light which comes from a broad light source and creates shadows that are not clearly defined.
Software	A computer program.
Standard lens	A lens that gives a view that is close to normal visual perception.
Steep perspective	Exaggerated diminishing perspective created by a viewpoint in close proximity to the subject with a wide-angle lens.
Stop down	Decreasing the aperture of the lens to reduce the exposure.
Straight photography	Photographic images that have not been manipulated.
Subjective analysis	Personal opinions or views concerning the perceived communication and aesthetic value of an image.
Symmetry	Duplication of information either side of a central line to give an image balance and harmony.
Sync lead	A lead from the camera to the flash unit which synchronises the firing of the flash and the opening of the shutter.
Sync speed	The fastest shutter speed available, for use with flash, on a camera with a focal plane shutter. If the sync speed of the camera is exceeded when using flash the image will not be fully exposed.
System software	Computer operating program, e.g. Windows or Mac OS.

Telephoto lens	A long focal length lens. Often used to photograph distant subjects which the photographer is unable to get close to. Also used to flatten apparent perspective and decrease depth of field.
Thematic images	A set of images with a unifying idea.
TIFF	Tagged Image File Format. Popular image file format for desktop publishing applications.
Tone	A tint of colour or shade of grey.
Transparent	Allowing light to pass through.
TTL meter	Through-the-lens reflective light meter. This is a convenient way to measure the brightness of a scene as the meter is behind the camera lens.
Tungsten light	A common type of electric light such as that produced by household bulbs and photographic lamps. An 80A blue filter may be used to prevent an orange cast.
Unsharp mask	A filter for increasing apparent sharpness of a digital image.
Uprating film	The action of 'pushing' the film, i.e. the film speed on the camera's dial is increased to a higher ISO for the entire film. This enables the film to be used in low light conditions. Development must be increased to compensate for the underexposure.
UV filter	A filter used to absorb ultraviolet radiation. The filter appears colourless and may be left on the lens permanently for protection.
Vantage point	A position in relation to the subject which enables the photographer to compose a good shot.
Vertical	At right angles to the horizontal plane.
Virtual memory	Hard drive memory allocated to function as RAM.
Visualise	To imagine how something will look once it has been completed.
Wide angle lens	A lens with an angle of view greater than 60°. Used when the photographer is unable to move further away or wishes to move closer to create steep perspective.
X	Synchronisation setting for electronic flash.
X-sync (pc socket)	A socket on the camera or flash unit which enables a sync lead to be attached. When this lead is connected the flash will fire in synchronisation with the shutter opening.
Zoom lens	A variable focal length lens. Zoom lenses have comparatively smaller maximum apertures than fixed focal length lenses.
Zooming	This is a technique where the focal length of a zoom lens is altered during a long exposure. The effect creates movement blur which radiates from the centre of the image.

Index